T5-ARJ-288

# THE NEW TELECOMMUNICATIONS

# THE NEW TELECOMMUNICATIONS

## A Political Economy of Network Evolution

Robin Mansell

SAGE Publications
London · Thousand Oaks · New Delhi

First published 1993

SAGE Publications Ltd
6 Bonhill Street
London EC2A 4PU

SAGE Publications Inc
2455 Teller Road
Thousand Oaks, California 91320

SAGE Publications India Pvt Ltd
32, M-Block Market
Greater Kailash – I
New Delhi 110 048

**British Library Cataloguing in Publication data**

Mansell, Robin
New Telecommunications: Political Economy of Network Evolution
I. Title
384
ISBN 0–8039–8535–5
ISBN 0–8039–8536–3 (pbk)

**Library of Congress catalog card number 93–085414**

Typeset by Type Study, Scarborough
Printed in Great Britain by Redwood Books, Trowbridge, Wiltshire

*In memory of Dallas Smythe*

Choice manifests itself in society in small increments and moment-to-moment decisions as well as in loud dramatic struggles; and he who does not see choice in the development of the machine merely betrays his incapacity to observe cumulative effects until they are bunched together so closely that they seem completely external and impersonal.

What we call, in its final results, 'the machine' was not . . . the passive by-product of technics itself, developing through small ingenuities and improvements and finally spreading over the entire field of social effort. On the contrary, the mechanical discipline and many of the primary inventions themselves were the result of deliberate effort to achieve a mechanical way of life: the motive in back of this was not technical efficiency but holiness, or power over other men. In the course of development machines have extended these aims and provided a physical vehicle for their fulfilment.

(Mumford, *Technics and Civilization* 1934: 6, 364)

# Contents

# Acknowledgements

The research on the intelligent network which forms the core of this study was supported by a project grant from 1989 to 1991 within the Economic and Social Research Council's Programme on Information and Communication Technologies in the United Kingdom. Numerous members of the industrial and policy community, particularly in the United Kingdom, France, Germany and Sweden, gave freely of their time during interviews and subsequent discussions connected with the research.

Colleagues at the Science Policy Research Unit, University of Sussex, contributed their enthusiasm and insights, and ensured that our research team continued to thrive despite my preoccupation with this book. Kevin Morgan, an important instigator of the project, moved to the University of Wales, Cardiff, but participated as colleague and friend throughout. Andrew Davies, now at the University of Amsterdam, and Susanne Schmidt, now at the Max-Planck-Institut in Köln, contributed in the middle stage by preparing interview materials, documentation and the draft text of parts of Chapters 4 and 9. In the final stage, Richard Hawkins and Michael Jenkins offered comments and editorial suggestions and Paul Walker prepared the graphics. Cynthia Robson added much-needed final touches to the manuscript. Natasha Constantelou, Ken Guy, Jane Millar, Paul Quintas, Roger Silverstone and Betty Skolnick also found themselves involved and I am very grateful for their support.

My perspective on the evolution of the public telecommunication network has been influenced over the years by numerous colleagues in many countries. I am especially grateful to Christopher Freeman, Richard Gabel, Nicholas Garnham, Richard Hawkins, Stuart Macdonald, William Melody, Kevin Morgan, Rohan Samarajiva, Liora Salter, Susanne Schmidt and Dimitri Ypsilanti for what they have taught me in their various ways and for never failing to challenge my thinking. Bill, Liora, Rohan, Stuart, Susanne, and Walter Bolter offered comments on the penultimate draft which were very helpful in strengthening the essential arguments. Finally, Sage Editorial Director Stephen Barr's, my mother's and Thomas Fawkes' certainty that this work would be completed provided the encouragement necessary to ensure that it has been. I am fully accountable for any errors or omissions and, of course, for the interpretation of events.

*R.M.*
*Science Policy Research Unit*
*University of Sussex*

# Introduction

The social and economic prospects of regions, countries, firms and individuals are linked to the evolution of the advanced telecommunication infrastructure. In the middle of the 1980s, the architectural concept for an *intelligent network* emerged in the United States and in Europe. The intelligent network specified a way in which computerized databases and complex software could be used to expand the capabilities of the telecommunication network. Intelligent networks were being implemented in more than thirty countries in 1993 and these accounted for nearly 80 percent of the world's telephone lines (Sleath 1993).

As work has continued on the design of the intelligent network, it has become clear that advanced information-processing services are increasingly important mediators of social, political and economic relationships. The intelligent network will underpin the provision of most communication services in one way or another in the coming years, and its capabilities are enabling the telecommunication network itself to become a marketplace for the exchange of goods and services. The intelligent network architecture is undergoing constant modification, and its development offers an opportunity to investigate how the public telecommunication network is being shaped by the dynamic forces of political and economic change in national, regional and global markets.

The embedding of greater intelligence in the public telecommunication network is expected by some forecasters to provide a basis for widespread access to public network services for all customers. However, at almost the same time as the intelligent network architecture was being fashioned in the laboratories of the telecommunication engineers, a report called the *Missing Link* was drawing attention to the fact that many people in developing countries had no access even to traditional telephone services. The telephone penetration rate – the number of main telephone lines per 100 people – in developing countries in the early 1980s was hovering around five in India, and less than one in 100 in some African countries (Independent Commission for World-wide Telecommunication Development 1984). The average penetration rate had reached about forty-three main telephone lines per 100 people in the OECD countries in 1990 and was only slightly lower than this in the European Community (OECD 1992). There was little improvement over the decade of the 1980s in the developing countries. Such disparities in access to the means of electronic communication are likely to grow, not only between industrialized and developing countries, but also within these countries. In the European Community, for example, the penetration rate for Germany as a whole in 1990 was thirty-nine main telephones per 100 people, but in the eastern regions within Germany, the rate was only ten.[1]

These figures do not indicate whether access is provided to the voice telephone network or to more advanced services, or whether available services are affordable. For most of the twentieth century, participation in the electronic communication environment has required access to telex and telephone services. Today, the exchange of social, cultural and business information often requires access to complex services based upon networks which combine advanced computing and telecommunication hardware and software. The current process of network evolution is giving rise to new terms and conditions of access to the public network and these may become increasingly divisive. The forces of technical, economic and political change in global markets are creating pressures to abandon traditional public interest concerns which have guided telecommunication policy in the industrialized countries.

Telecommunication policies, at least in theory, have been devised to ensure that the public network enables smaller and larger firms and residential consumers to make telephone calls and to use more advanced services as they become available. Although access to the public network provides no guarantee that economic resources and skills will be available to enable users to take advantage of these services,[2] the terms and conditions of access are instrumental in determining who can participate fully in the social, cultural, political and economic life of society. The public telecommunication network is being designed in an environment that is influenced increasingly by the competitive prospects and priorities of a relatively small number of privileged, oligopolistic firms. An important issue for public policy is whether the evolution of advanced public networks will extend access to new services or create new barriers which will render these services accessible mainly to the most privileged customers. This study examines whether present policies and regulatory institutions can ensure that a broader interpretation of the public interest in network evolution informs the design and implementation of the intelligent network.

The terms and conditions of public network access are influenced by technical standards, restrictions on the use of networks and the prices of services. These aspects are shaped by the political and economic priorities of a variety of public and private institutions. The political economy perspective which is developed in this study focuses upon how, and by whom, these priorities are being established and the consequences for the accessibility of the electronic communication environment of the twenty-first century. The analysis is centred upon the strategies of public telecommunication operators, equipment manufacturers, and the multinational users of telecommunication services in the late 1980s and early 1990s in the United States and in Europe. Two clearly discernible clusters of alternative views of how the technical innovation process in the telecommunication industry interacts with political and economic forces are explored. One perspective sees the development and diffusion of technical innovations, such as the intelligent network, as resulting in flourishing competition and a declining need for regulation. The other envisages a transformation from the monopolistic telecommunication supply arrangements of the past to oligopolistic market structures in which global rivalries increasingly set the priorities for public network development. The result is a continuing need for regulation to safeguard the public interest in access to the public telecommunication network.

These two perspectives provide a framework to explore the implications of the strategies and tactics of the telecommunication supplier, user and policy community. Documentary and interview materials are used to illustrate how the political and economic forces of change are intertwined with technical developments. The architecture of the intelligent network is shown to be malleable and informed by a variety of political and economic factors. The analysis considers the origin of the intelligent network architecture in the United States at a time of major structural change. The divestiture of AT&T in 1984 created new challenges for the regulation of public telecommunication operators. Changes in the policy and regulatory environment were also gathering force in Europe. The liberalization of domestic telecommunication markets in the United States and Europe provides the context for a detailed exploration of the major determinants of the recent history of public network evolution.

Case studies of the engineering and marketing perspectives of the public telecommunication operators and major equipment suppliers in France, Germany, Sweden and the United Kingdom are used to show how the terms and conditions of access to the public network are being established, often in ways that elude the scrutiny of public policy institutions. Many representatives of public and private organizations contributed their views on the design requirements for the intelligent network and the appropriate role of regulation. These contributors are not named individually and their views are attributed to the institutions they represent. However, the nuances of their disparate, and often conflicting, perspectives are highlighted by differentiating between views informed by an idealistic vision of the process of technical and institutional change and those informed by the strategic priorities of decision-makers in the public and private sphere.

The intelligent network's evolution has profound implications for the future design of the public telecommunication network. Public policy, whether through state provision of telecommunication services or through public regulation of private monopolies, has been necessary to promote access to basic telephone services, although the success of such policies has varied considerably. The accessibility of some of the advanced intelligent network services must be encouraged if increasing numbers of smaller firms and residential consumers are not to be excluded from the electronic communication environment in the coming decades. The global telecommunication networks of the 1990s respect few political boundaries, and the resulting challenge to public policy and regulation is examined for its implications at the national and regional levels in Europe.

The technical and the institutional innovations in the public telecommunication network are investigated in this study. The technical aspects of network evolution are not regarded as holding solutions to the disparities in the accessibility of communication services. Nevertheless, the technical domain is a site where priorities for network development trajectories are forged and the technical developments, together with the political and economic factors that inform them, show that the intelligent network which is promoted in the rhetoric of the supply side of the industry differs in some very important ways from that which is actually being built. When this is understood, realistic goals for telecommunication policy can be developed and opportunities for more effective regulatory intervention in the telecommunication market can be devised.

## Notes

1 In Italy, an overall penetration rate of thirty-nine main telephones per 100 people could be compared with a rate of only twenty-nine in the Less Favoured Regions of that country. In Britain, an overall penetration rate of forty-three main telephones camouflaged the fact that there were only thirty-three main telephones per 100 people in the designated Less Favoured Regions. Penetration rates are for main lines per 100 inhabitants at the country level and the Objective 1 (Less Favoured Region) level within countries (Ewbank Preece Ltd 1993).

2 'Access' is a difficult construct to define and the term is used in a variety of technical and non-technical ways in this study. For a review of the diverse ways in which 'access' has been considered in relation to concepts of 'public service', 'universal service' and 'technical access', see Garnham and Mansell 1991; Jacobson 1989; Mansell 1989.

# 1

# The Biased Structuring of Telecommunication Networks

> Innovation is of importance . . . It enables the whole quality of life to be changed for better or for worse. It can mean not merely more of the same goods but a pattern of goods and services which has not previously existed, except in the imagination.
>
> (Freeman 1982: 3)

## Telecommunication Policy for a Networked Economy

Access to telecommunication services is becoming one of the minimum necessary conditions for participation in domestic and international markets. Historically, as Braudel has suggested, the effective exploitation of commercial trading networks depended upon supporting infrastructures such as ocean navigation, canals and roads. For Braudel, the pace and ease of movement within a network was conditioned, not by the maximum potential of any one part of the network, but by the minimum characteristics of the individual elements present across the network (Braudel 1984). Similarly, for the telecommunication network, its role in the economy and society is a reflection of the minimum characteristics that are available to all those who would benefit from its use.

The public telecommunication network is an important medium for communication and, as such, its design and availability shape the conduct of social and economic life. Although the causal linkages between investment in telecommunication networks and social and economic development are difficult to establish, there is considerable evidence that political and economic control within society is contingent upon the characteristics of the electronic communication environment (Bressand 1988; Cruise-O'Brien 1983; Robinson 1991).[1] If telecommunication networks are regarded as being analogous to the *nervous system* of society,[2] then, in principle, they should be all-pervasive, just as oral and paper-based networks of communication have been.

For some observers, the evolution of public telecommunication networks raises issues that do not differ from those that need to be considered in the manufacture of shoes or automobiles. Nevertheless, both the social and economic implications of recent trends in the structure and organization of the public telecommunication infrastructure have become major issues in the 1990s. The telecommunication service industry contributed slightly more than 2.25 percent to the Gross National Product in the OECD area in 1990. In the OECD countries in 1990 investment in public telecommunication networks reached US$85 billion (OECD 1992). This is a major industry, and its prospects have captured the attention of the policy community, the globally operating firms that supply and use advanced telecommunication networks, and disparate coalitions of smaller firms and individual consumers in the United States and, increasingly, in Europe.

Policy analysis generally is able to map the scientific conception of an original technical idea such as the intelligent network. Studies of the gestation of an invention often point to the creative role played by scientists, engineers and their supporting institutions. Such studies also focus on characteristics of the corporate and public policy environments that are conducive to sustainable systems of innovation. However, policy analysis turns less often to examine factors which contribute to an invention's subsequent innovation and diffusion. When it does so, analysis often focuses upon the more tangible elements of the process of research, development and commercial implementation such as expenditure, patenting activity, sales revenues and other indicators. However, public policy-makers, corporate strategists and the policy research community would do well to heed the words of T. S. Eliot: 'between the idea and the reality, between the motion and the act . . . between the conception and the creation . . . falls the shadow' (Eliot 1969). Neglected in the shadows of much policy research in the telecommunication field are fundamental issues associated with the myths and realities of the dynamics of technical and institutional change.

In this study, the political and economic factors which are shaping the conception and development of intelligent networks are the main concern. 'Intelligence' refers to the capacity of the telecommunication network to process, route and store information using digital switches and transmission links, computers and databases. In an intelligent network, a substantially greater degree of information-processing capability resides within the public network as compared to the traditional capabilities of the telecommunication infrastructure. The public telecommunication infrastructure is expected to evolve to offer an intelligent platform which can be used to supply advanced information and communication services. These services involve information-processing capabilities far beyond those needed for the simple conveyance of messages.

Telecommunication policies for the networked economy will need to be grounded by policy analysis that goes beyond aggregate statistical indicators of the innovation and diffusion of technical systems and the performance of firms involved in their production. Investigation of the factors shaping the contours, dynamics and biases of evolving telecommunication markets is needed.[3] Many of the most controversial policy issues in this area concern disparities in the conditions of access to the electronic communication environment. Although the existence of disparities is often acknowledged, this rarely leads to exposure of the determinants of uneven network development. To accomplish this, it is necessary to investigate the underlying political and economic determinants of technical and institutional change in the telecommunication industry.

## Advanced Telecommunication Network Evolution

There was a major technical shift in the telecommunication network in the 1980s from analogue to digital switching and transmission. There was a quantum leap from limited computerized control of electronic switching towards increased reliance upon the intelligent functions embodied within the telecommunication network. Software-based computerized functionality now supports the electronic processing of digital signals, and this capability is often referred to as the intelligent telecommunication network. These software-based functions extend from network management to advanced service applications.

Alongside this technical shift there was a policy shift. Competition was embraced as a superior means of organizing telecommunication supply as compared with the public and private monopolies of the past. Policy shifted towards encouraging the mixing and matching of switched and dedicated systems, and a blurring of distinctions between public and private networks began to emerge. A *network of networks* came to be seen as a means of responding to the diverse requirements of multinational businesses, small and medium-sized firms and the mass consumer market.[4] In some quarters, this shift led to advocacy of the withdrawal of public intervention in the market for telecommunication network and service supply. The competitive marketplace was, and continues to be, championed as the mechanism best able to ensure that advanced telecommunication networks contribute to competitiveness, improved productivity, efficiency and widespread service diffusion.

Behind these technical and policy shifts lies a complex vision of an information economy in which advanced communication services supported by intelligent telecommunication networks become far more significant generators of economic wealth than the traditional telecommunication service sector.[5] Today, the public telecommunication operators (PTOs) are seeking to develop a position of strength from which to attack these future markets. They are being challenged by service suppliers with experience in producing, processing and distributing information and communication services. The telecommunication equipment manufacturers also face challenges to their markets from computer manufacturer and software developer firms.

Telecommunication is an industry at the centre of the all-pervasive shift in techno-economic paradigm that Freeman and others regard as the progenitor of widespread institutional change. The term 'techno-economic paradigm' refers to 'a combination of interrelated product and process, technical, organisational and managerial innovations embodying a quantum jump in potential productivity for all or most of the economy and opening up an unusually wide range of investment and profit opportunities' (Freeman and Perez 1988: 47–8; see also Perez 1983). The evolution of intelligent telecommunication networks provides an opportunity to investigate the political and economic forces embodied in the institutions that are shaping technical outcomes in the telecommunication field.

The evolution of the intelligent network has been discussed in the trade and technical literature as a network architecture with a set of attributes that can be accepted or rejected on the basis of technical criteria (Begbie et al. 1982; Eske-Crisstensen et al. 1989; Gilhooly 1987; Gilhooly 1991b). However, if the implications of network development are to be adequately understood, policy analysis must be more broadly concerned with the political and economic objectives that are being embedded in this innovative technical architecture. These objectives are influential determinants of whether alternative network designs are accepted or rejected. The study of the political and economic determinants of technical and institutional change in the communication field has thus far tended to be concerned retrospectively with the factors which have led to the structure of markets, policies and regulatory institutions (Babe 1990; Gabel 1967; Melody 1986; Smythe 1957; Trebing 1969a).

Few studies have focused on the current processes of telecommunication network evolution as an instance of technical and institutional innovation. The pervasiveness of the electronic communication environment is such that the reverberations of changes in the 1980s and early 1990s in the telecommunication industry are being felt across all facets of socio-economic activity. Seemingly neutral technical decisions with respect to the design of the intelligent network are forging the contours of the re-distribution of market power within, and beyond, the traditional telecommunication industry. Rivalry among the major firms on the supply and demand side of the industry is contributing to the process of *monopolization* (Clark 1961). The implications of the new telecommunication environment also need to be considered from the perspective of the users of public telecommunication networks. These users, or 'customers' as the industry has begun to call them, have extremely heterogeneous requirements for services that can be supported by intelligent networks.

The transformations in the structure and organization of telecommunication markets have been examined over the past decade from many vantage points, including the contribution of services to the economy and society, the changing structure and purpose of regulation, the process of service innovation, and the restructuring of national and international telecommunication policy institutions.[6] These investigations have broadened debate on the implications of the changing structure of telecommunication. However, they have often contributed to the myth that digital information-processing technologies are impelling the public telecommunication network towards a future in which competition flourishes to the benefit of all users. Many of these studies consider how suppliers, users and policy-makers are adjusting to the diffusion of advanced communication technologies. The ways in which political and economic factors and existing institutions give rise, or give way, to innovations in technical artefacts in the telecommunication industry, thereby constraining and structuring their use, have barely been considered. Nevertheless, it is these relationships that need investigation if we are to understand how technical design is mediated by political and economic power. As Smythe observed, 'science and technique necessarily involve *choices* of problems to be studied and knowledge to be put into practice and . . . such choices arise out of and are conditioned by, as well as affecting in turn, the on-going social structure of power relations' (Smythe 1978).

Technical studies of engineering choices, network architectures and standards in the telecommunication field have rarely been linked with analyses of social, economic or political change.[7] The intelligent network has been treated as the next step in the development of advanced communication services that will inevitably bring economy- and society-wide benefits (Ambrosch et al. 1989; International Resource Development 1990). However, there is much controversy over the appropriate technical design for the intelligent network. Although the architecture for the intelligent network was first announced in the mid-1980s, almost a decade has passed before signs of vigorous public debate on its implications have begun to emerge. For example, it was not until 1992 that the Department of Trade and Industry in Britain launched a discussion document to stimulate debate (Department of Trade and Industry 1992). A response by an influential telecommunication user organization provided a glimpse of the importance of the non-technical issues: 'no one operator should be allowed to gain a stranglehold over the Intelligent Network, especially its development, distribution and access' (Telecommunication Managers Association 1993: 16). In this study, the recent history of the technical design of the intelligent network is used as a guide to future technical and institutional developments. The competitiveness of the telecommunication industry and the benefits of advanced services have been addressed in the economics, management and trade in services literature (Antonelli 1985; Ciborra 1992; Granstrand and Sjolander 1990; Pisano et al. 1988; Ungerer and Costello 1988). The predominant concern with the international competitiveness of suppliers and larger users is symptomatic of significant biases in the way network access conditions are developing in the global networked economy.

## Contrasting Views of Global Telecommunication Development

To assist the analysis of alternative views of trends in telecommunication development, two alternative models are considered. The first is referred to as the *Idealist* model. This model is derived from theories which envisage the emergence of a mature and fully articulated competitive market. In this model, the following characteristics of the market are present: a large number of sellers are active in the market for goods and services such that the impact of one seller on the market is negligible; buyers perceive that sellers produce a homogeneous product; and buyers have access to perfect knowledge, or at least sufficient knowledge, to enable them to make informed, rational decisions (Clark 1961). This type of competition is characterized by the absence of barriers to market entry and exit. This model falls within the genre of idealistic analyses which are based upon 'a mystified and mystifying notion of technology as a cure for the world's very real problems' (Smythe and Dinh 1983: 120).

The second model is referred to as the *Strategic* model. It is rooted in theories of imperfect competition, monopolistic competition (competition among differentiated products), oligopolistic rivalry (competition among small numbers of suppliers) and monopoly.[8] This model is one which seeks to provide a 'reality-based analysis of institutional processes' (Smythe and Dinh 1983: 120). In this model, it is assumed that institutions are characterized by 'indeterminate, unstable oligopoly wherein the transnational corporations deliberately employ short-run pricing strategy to achieve long-run entrenchment and monopoly power in national markets, foreign and domestic' (Melody 1985). The structure of markets is enmeshed with technical change and the determinants of change are located within a broad array of social and institutional arrangements. They are not found solely within the price mechanism and the effects of exogenous shocks created by technical change. As Freeman has observed:

> The socio-institutional framework always influences and may sometimes facilitate and sometimes retard processes of technical and structural change, coordination and dynamic adjustment. Such acceleration and retardation effects relate not simply to market 'imperfections', but to the nature of markets themselves, and to the behaviour of agents (that is, institutions are an inseparable part of the way markets work). (Freeman 1988)

The prevailing vision of global telecommunication development is one which incorporates the assumptions of the Idealist model. At the root of this model is the notion that a single trajectory of development will prove to be inherently superior and will come to be reflected in the technical composition of the advanced telecommunication infrastructure.[9] In so far as the evolutionary process of technical innovation in telecommunication and computing technologies is left to the forces of the competitive market, the network will tend to embrace the collective interests of policy-makers, suppliers and all users.

The Idealist model stands in stark contrast to Mumford's observation that 'technics and civilization as a whole are the result of human choices and aptitudes and strivings, deliberate as well as unconscious, often irrational when apparently they are most objective and scientific' (Mumford 1934). In contention with the Idealist model, as a more realistic account of the strategic and tactical manoeuvring of multinational firms and the political and economic determinants of policy and regulatory regimes, is the Strategic model. In the Strategic model there is continuous rivalry among a relatively small number of dominant firms. Rivalry, monopolization and institutional restructuring do not serve all market participants equally well, but neither do they preclude the presence of smaller firms which sell a relatively narrow range of goods or services. Rivalry among suppliers and telecommunication users creates the impetus for dynamic changes in public and private institutions. The Strategic model emphasizes the ways in which new market distortions are created and become embedded in the design of technical artefacts such as the intelligent network. In this model, the need for policies to ameliorate outcomes deemed to be socially, politically or economically unacceptable does not fade away.

### *The Idealist Model*

The Idealist model of telecommunication development incorporates the following assumptions with regard to the evolution of the intelligent network:

1. Intelligent networks will provide the basis for the integration of information and communication services within a permeable seamless network. Boundaries between public and private networks will disappear and the convergence of competencies across the telecommunication, computing (and audio-visual) sectors will abolish the distinctive core competencies which divided these sectors in the past.
2. Intelligent networks will provide the basis for ubiquitous or universal service diffusion. Technical (software) and organizational constraints will be overcome partly as a result of the declining costs of extending services throughout the network and partly as a result of the feasibility of responding almost instantaneously to customer requirements expressed via interactive network services.
3. Intelligent networks will transform the telecommunication industry from a supply-led to a demand-led industry like other producing sectors of the economy which are assumed to perform according to the rules of a fully competitive market. The design and implementation of the intelligent network will necessitate this shift.
4. The shift to a demand-led industry will force a rationalization of the supply side and stimulate the introduction of flexible, high-quality services at reduced cost. New entrants in equipment, network operator and service supplier markets – both foreign and domestic – will erode the market power of incumbents and stem the rise of new dominant players. Any residual imperfections in the market can be addressed via minimal forms of regulation, e.g. the introduction of open systems and common standards.
5. The intelligent network will stimulate new forms of collaboration and competition which arise from the creative stimulus of the supply–demand side balance. Collaboration will result in benign partnerships that are responsive to all facets of consumer demand. Competition will permeate telecommunication supply from equipment, to local, national,

regional and global service supply, ensuring ubiquitous access to advanced networks and services in the long term.

6 During a transitional phase before the intelligent network becomes fully diffused, specialized telecommunication regulatory institutions at the national and regional level will ensure that intelligent network development evolves on a trajectory that ensures that efficiency and public service objectives are met.

The Idealist model assumes that the combined forces of technical innovation and competition will erode monopolistic control of the telecommunication infrastructure and the services it supports. In this model, the PTOs argue successfully that the diffusion of an advanced telecommunication infrastructure and the entry of new service providers means that no single supplier can dominate the market sufficiently to foreclose entry or to discriminate unjustifiably among customers. Telecommunication supply is treated like any other competitive commercial activity. The PTOs are not constrained from entering any segment of the telecommunication market. New entrants are not protected from the strategies and tactics of the dominant incumbents in the market. The PTOs welcome competition and market liberalization since they are free to compete across the same range of services as their competitors. They are free to offer differentiated services and prices to their customers and to relinquish their implicit or explicit public service obligations. Although there is a period of transition as the forces of effective competition take hold, in certain countries the transition is largely completed. In countries where the diffusion of the advanced telecommunication infrastructure has been slower, competition provides the best incentive for stimulating a 'catch-up' or 'leap-frogging' of earlier phases of telecommunication development.

Under the Idealist model, it is assumed that the PTOs are fully answerable to all their customers. The intelligent network is seen as an engineering marvel which can provide the key to ubiquitous advanced services and to fair competition. It can provide a technical solution to the need to provide an electronic communication environment which contributes to the competitiveness of the economy and to the political, social and cultural cohesion of markets and communities. The need for policy intervention and regulation is fading away and should be replaced by a referee who arbitrates small disputes as they arise.

Within the Idealist model of telecommunication development, the intelligent network is treated as a direct response to customer requirements. Although not all customers have a voice in the marketplace, the largest customers are instrumental in shaping investment and network design decisions in the interests of all users. The development of the intelligent network is expected to prise open the public telecommunication infrastructure and to vanquish the monopoly power of the PTOs. Moreover, it is expected to bring to a halt the monopolization of the telecommunication industry.

*The Strategic Model*

The Strategic model is premised on the following propositions for the evolution of the intelligent network:

1 The economic and political interests that become embedded in the design and implementation of the intelligent network will be unlikely to lead to network integration and a seamless global intelligent network infrastructure. Rivalry expressed through pricing, proprietary standards and policy intervention will create new disparities in the process of network evolution.
2 As a result of market imperfections, it is unlikely that there will be a ubiquitous diffusion of advanced communication services and there will be disparities and an uneven development of the terms and conditions of network access.
3 The design of the intelligent network will be largely supply-led. Where it is subject to pressures on the demand side, these will reflect primarily multinational business-communication requirements.
4 Technical innovations in the intelligent network (and by analogy in other network configurations) will provide a weak stimulus for competition, and this will be insufficient to prevent monopolization within the industry.
5 The balance between the supply- and demand-side forces in the development of the intelligent network will create incentives to design networks that help to maintain or re-establish monopolistic power in the marketplace. Network segmentation, i.e. the mixing and matching of network components, will result in flexibility and choice for multinational companies but it will produce dis-benefits for many other consumers in terms of network access costs and more complex network access conditions.
6 New regulatory institutions at national and regional levels will introduce pressures that stimulate telecommunication suppliers to devise new ways to maintain market power. Telecommunication network development will display greater efficiencies in some segments of the industry, but little progress will be made towards public service objectives for advanced services when compared with the ubiquity of the public telephone service in most industrialized countries.

The Idealist model insists upon the effectiveness of competition and technical change in eroding monopolistic control of telecommunication development. Its counterpart, the Strategic model, draws upon theories of institutional change and market power to determine the modalities through which such power is executed. Applied to the intelligent network, this model shows the strengths and weaknesses of the forces of competition. Where competition does exist, it displays signs of superficial product differentiation, effective service competition only in certain submarkets, and closure of network access at key interfaces in the network infrastructure.

The premises of the Strategic model are visible in the views of suppliers and large telecommunication users when they find themselves confronted by actual distortions in the marketplace. For example, Allen, Chairman of AT&T, has observed that, in reality, the market does not conform to the tenets of the Idealist model.

> The United States experience has also taught us that competition in long distance is virtually impossible if one of the competing carriers also has monopoly control of the local exchange service. Any company that has a monopoly over local service has the power to subvert competition in long distance [markets], domestic or international. (Rhodes 1991)

The willingness of those who appear to favour the Idealist model of telecommunication development to acknowledge inequalities in the distribution of market power is generally a function of their perceptions of the need for strategic positioning within dynamically changing markets. In circumstances where the dominant actors are challenged, they often argue that the market would acquire the attributes of the Idealist model if public intervention in the market, and regulatory restraints, were removed.

Since the mid-1980s, a realignment of telecommunication network access and control among PTOs, equipment manufacturers, and the large telecommunication users has been underway. The result could be a public network in which the design of the intelligent network gives smaller businesses and residential customers even less control over the electronic communication environment than they had under the old monopolistic regime.

In the cases of both the Idealist and the Strategic models, the assumptions outlined above can be relaxed. For example, it has been argued by proponents of the Idealist model that markets are competitive if they are contestable; that is, when a dominant or monopoly firm perceives and reacts to a challenge to its position in the market. The perceived challenge can be based upon actual entry or upon expectations concerning changes in the regulatory environment that could strengthen the potential for entry by reducing legal and other barriers (Baumol et al. 1982). Similarly, proponents of the Strategic model use a variety of indicators to determine whether companies hold a dominant position in the market. They are not always bound by the same determination of, for example, the share of a 'relevant market' that is sufficient to enable a company to exercise its market power (Scherer 1980).

However, as Clark commented with respect to theories of competition, monopoly and oligopolistic rivalry, 'it is a commonplace that the facts depart from any feasible theoretical formulations, and the latter are supposed to serve, not as letter-perfect descriptions of the facts, but as points of departure for understanding them' (Clark 1961). It is in this spirit that the Idealist and Strategic models are used to highlight the polarization of views on the implications of the evolution of intelligent networks.

**Introducing the Intelligent Network Strategists**

The articulation of rivalry and monopolization in the telecommunication industry is examined through the lens of the design choices that are embedded in the development of the intelligent network. In the United States, the study focuses mainly on AT&T and Northern Telecom, the Canadian-owned equipment manufacturer. In Europe, developments in the United Kingdom are examined through an analysis of British Telecom and Mercury together with GEC Plessey Telecommunications (GPT) and Ericsson UK. In France, the study focuses on France Télécom and Alcatel; in Germany, on Deutsche Bundespost Telekom (DBP Telekom), Siemens and Standard Elektrik Lorenz (SEL Alcatel); and in Sweden, on Swedish Telecom (Televerket) and Ericsson.

In 1990–91 the equipment sales of the telecommunication manufacturers in this study accounted for 67 percent of the sales of the top fifteen telecommunication equipment manufacturers in the world.[10] AT&T was in the lead with US$13.3 billion, followed by Alcatel with $11.6 billion, Siemens with $9.9 billion, Northern Telecom with $7.7 billion, Ericsson with $7.1 billion and GPT with $2.2 billion (OECD 1992).

These companies produce the telecommunication switches which make changes in the connections between both analogue and digital circuits. Originally comprised of electromechanical devices made of rotary switches (selectors), the automatic telephone exchange was invented in 1889 by Strowger. An improved crossbar exchange was introduced in the 1920s and the design principles were later adapted for early electronic switches. Digital computers can be used to control electronic switches; for example, Stored Program Control (SPC). Time, frequency and code division multiplexing systems are being developed, and these require digital transmission capability. These companies also supply Private Branch Exchanges (PBX) or Private Automated Branch Exchanges (PABX) that are usually located on a company's premises and function as switched systems which provide services within premises. This arrangement can be extended by leased circuit connections and interworking with the public switched telecommunication network to extend private networks to multiple sites and between firms whose operations often span the globe. PBXs are designed primarily to support voice communication.[11]

The European-owned equipment manufacturers, including Alcatel, Siemens and Ericsson, are aspiring to retain or improve their ranking in the world telecommunication equipment market in the face of challenges from companies such as AT&T and Northern Telecom as well as from the computer manufacturers with IBM, DEC and Tandem Computers playing a significant role. The telecommunication equipment trade balance in the United States was negative at US$ −1.9 billion in 1990, as was the trade balance for the United Kingdom, at $ −223 million. Both had moved from a position of surplus in 1980. The other three countries had remained in a position of surplus throughout the 1980s and, by 1990, there were positive trade balances of $657 million, $1.0 billion, and $1.3 billion for France, Germany and Sweden, respectively (OECD 1992).

AT&T, MCI, Sprint and a variety of international value-added service suppliers have been challenging the supremacy of the European nationally based PTOs. The telecommunication service market generated revenues of US$367 billion for the public telecommunication operators in the OECD countries in 1990. The US market was responsible for 47 percent of this total, and France, Germany, Sweden and the United Kingdom, for 20 percent. AT&T's revenues from public telecommunication services in 1990–91 were US$20.4 billion. The largest PTOs by revenues in the European Community in 1990–91 were Deutsche Bundespost Telekom with revenues of $25.1 billion, British Telecom with $23.4 billion, and France Télécom with $21.1 billion. Swedish Telecom, with $5.7 billion, although a smaller operator, provides a basis for comparison with the other major companies in the European market (OECD 1992).

In the United States, the penetration rate for main telephone lines in 1990 reached forty-five per 100 inhabitants. The penetration rates in the four European countries were fifty, forty-seven, sixty-eight, and forty-four for France, Germany, Sweden and the United Kingdom, respectively (OECD 1992), and the voice telephone market was believed to be reaching saturation point.[12]

All the major operators in these countries had begun to outline strategies for generating revenues from advanced services in the mid-1980s when the intelligent network architecture emerged. These PTOs are aiming to become global, or at least regional, operators. They are upgrading the public telecommunication network at considerable cost and the intelligent network architecture plays a significant part in this modernization process. The PTOs are seeking to carry their multinational customers' voice, data and image traffic on their own networks. They do not want to be bypassed by competitors who could siphon off future revenues. The PTOs have plans to stimulate revenue growth by providing advanced information and communication services to the broader customer base – for example, the small and medium sized firms and the residential customer – but the primary target is increasingly the requirements of a relatively small number of globally operating firms.[13]

## The Intelligent Network Challenge – Opening the Black Box

The intelligent network may be an engineering marvel in the eyes of some observers, but it is not a key to a ubiquitous, seamless global infrastructure that will eradicate disparities in the terms and conditions of access to public networks. The intelligent network is one cornerstone of an innovation process whereby strategic advantage is continuously being created in a very imperfect marketplace. It is a design configuration that the PTOs can use to maintain their monopolistic power and the traditional telecommunication equipment manufacturers can use to resist erosion of their market share. It is also a tool whereby new entrants in the telecommunication market can seek to achieve greater control of the evolution of the public telecommunication network.

The processes of technical change, innovation and institutional restructuring embody technical and non-technical factors (Freeman 1982). Political, economic, social and cultural factors are embedded in the design and implementation of intelligent networks, and the story of how this occurs is a necessary complement to studies of corporate strategy, policy and regulatory reform that treat the telecommunication network as a technical black box (Mansell 1990). In the absence of analysis of the complex factors contributing to technical design, corporate strategies tend to blur into processes of tactical manoeuvring and uncertainty – a spectacle which makes it difficult to formulate policy or to build more effective institutions for regulation.

## Towards a Political Economy of Telecommunication

There is growing potential for the use of technical design by dominant suppliers in the telecommunication market as a mechanism which can help to elude the scrutiny of competitors and policy makers. If the Idealist model is the more accurate reflection of the realities of the telecommunication market, the competitive marketplace will ensure that network evolution brings benefits to all. However, if the Strategic model better represents reality, the technical design of public networks should not be left to those with acknowledged engineering expertise. The political economy of intelligent network design in the chapters which follow shows how design parameters affect the terms and conditions of network access and, ultimately, participation in the networked economy. As Innis remarked, 'a medium of communication has an important influence on the dissemination of knowledge over space and over time' (1951: 33). If policy and regulation are to encourage more equitable access to electronic means of communication, the social and economic issues raised by the technical design and implementation of the intelligent network must be addressed by a community far wider than the network engineers.

## Notes

1 For example, the viability of democratic processes is linked closely with the ubiquity of an advanced communication infrastructure (Garnham and Mansell 1991). The literature on the causal linkages between investment in telecommunication infrastructure and economic development has generally failed to show conclusive evidence of a direct causal linkage. See, e.g. Antonelli 1991; Cronin et al. 1991; International Telecommunication Union Advisory Group on Telecommunication Policy 1989; Saunders et al. 1983.

2 E.g., Gillick and Gilhooly (1990) have commented that 'telecommunications is now the

central nervous system of a service based global economy based increasingly on the exchange of electronic information. Electronic information exchange is the commodity that will fuel economic growth into the 21st century'; see also Commission of the European Communities 1987.

3 The 'global networked economy' is a label which has been closely associated with the rise of complex organizational structures which rely increasingly on advanced information and communication technologies, telecommunication networks and information services to support their business activities. See, e.g. Bressand 1990; Lanvin 1991.

4 The requirements or needs of telecommunication network users should be understood broadly to encompass the user's relationship with network artefacts themselves – e.g. terminal equipment, software functionality and so on – as well as the symbolic use and exchange value associated with these artefacts. This approach renders it virtually impossible to extricate the socio-economic conditions which shape the production of the telecommunication network from those which condition its use for the acquisition of information and the processes of communication. This approach follows Jhally's observation that 'the recognition of the fundamentally symbolic aspect of people's use of things must be the minimum starting point for a discourse that concerns objects', see Jhally 1990; Leiss 1976.

5 The 'traditional' telecommunication service sector refers here to revenues generated by the telex and telegraphy network, voice and data transmission services using the public switched telephone and data networks, as well as revenues from leased lines provided by public telecommunication operators.

6 See, e.g., the contributions of telecommunication services to the economy and society (Mulgan 1991); studies of the structure of regulation (Bruce et al. 1986; Hills 1986; Temin and Galambos 1987), the impact of telecommunication-related service innovations (Thomas and Miles 1990); studies of national and international policy institutions (Aronson and Cowhey 1988; Mueller and Foreman-Peck 1988).

7 An exception is found in David and Bunn's (1988) analysis of related issues in the electricity supply industry.

8 See, e.g. Clark's discussion of alternatives to the static competitive market model (1961); see also Schumpeter 1954. For a discussion of perspectives on the role of entry and exit barriers under conditions of imperfect competition in telecommunication, see Shepherd 1984.

9 This assumption contrasts sharply with David and Bunn's (1988) observation that 'where network technologies are involved, one cannot justifiably suppose that the system which has evolved is really superior to others which might have been developed further, but were not'.

10 This includes NEC Corporation, Bosch Telecom, Motorola, Fujitsu, Philips, Ascom, Italtel, Nokia and Matra.

11 The main public network switching systems produced by these suppliers are: the 5ESS – AT&T; DMS – Northern Telecom; AXE – Ericsson; EWSD – Siemens; System 12, E–10 – Alcatel; and System X – GPT which is now partly owned by Siemens.

12 The number of main lines per 100 population in Sweden in 1988 was 65, higher than any European Community member state. In the same year, the equivalent figures for the United Kingdom, West Germany and France were 40, 45 and 45 per 100 population, respectively.

13 Although it is not bound by them, Sweden has sought to ensure that its telecommunication equipment manufacturing and service market meets the terms and conditions of directives implemented by the European Community, especially concerning entry and use of public network facilities.

# 2

# The Intelligent Network – Changing Technologies and Institutions

> If you really want to be in business at the turn of the century as a telecommunications organisation . . . then the intelligent network is simply a revenue deployment strategy.
>
> (Bar and Borrus 1989: 3, citing Richard Snelling, BellSouth)

## A Turning Point in Network Evolution

In Europe national PTOs have moved gradually away from procurement of equipment based upon closed relationships with the traditional telecommunication manufacturers. One effect of the opening of the procurement process to competitive tendering in the late 1980s and early 1990s was the emergence of the computing industry as a potential source of expertise. When forging their business strategies the PTOs have been aware of the need to manage the technical convergence of telecommunication and computing in order to preserve their strength in voice telephone markets – and in potentially lucrative data and image markets.

Since the advent of computerized switching technologies,[1] distinctions between the computing and telecommunication industries have become focal points for controversy over the transformation of the telecommunication infrastructure into a flexible telematics platform upon which the *global information fabric* would come to rest (Rutkowski 1988). During the early collaboration between firms on both sides of this technical divide it was expected that such alliances would eventually produce optimal network designs (Solomon 1991). It was assumed that collaboration would be a relatively straightforward matter. Strategic management theory suggested that success was a matter of experimentation, learning and organizational change (Pisano et al. 1988). This theoretical perspective tended to assume that the exercise of market power would no longer be associated with the emergence of new disparities in the terms and conditions of network access. Instead, the exercise of market power would contribute positively to the strategic need on the part of globally operating firms to establish themselves in the international market.

However, it has not been these advanced technologies alone that have galvanized the telecommunication industry into a period of major upheaval in Europe. Nor are changes in the industry simply the result of cumulative learning experiences. Rather, economic and political incentives embedded in technical designs have contributed to, and shaped, the continuous restructuring of the industry. In so doing, they have mediated the changing power relations within the European market as well as beyond it. While technical innovation is a necessary precondition for change, it is the appropriation and control of innovations in technical artefacts which hold the keys to the direction of changes in the marketplace. The turning points in this process are visible in the struggle to gain control over the intelligent network.

The electro-mechanical telecommunication infrastructure was constructed by national PTOs with a distinctive organizational structure which remained intact for nearly fifty years. Public telecommunication networks had been developed mainly by public monopolies in the western industrialized countries, after an initial period of private entrepreneurial competition. In each country telecommunication equipment had been manufactured by a few firms protected from competition by the exclusive purchasing practices of the PTOs. This distinctive institutional structure has been described as a club or cartel (Cawson et al. 1990). Cawson et al. argue that the tensions between competition and closure led to a situation in which state power was co-opted to protect domestic telecommunication suppliers from the threat of new entry. A closed institutional structure with a few sellers (oligopoly) and a single buyer (monopsony) emerged.

This telecommunication club was instrumental between the 1920s and 1970s in the development of important technical innovations. During this period, a variety of incremental innovations resulted in improvements in the quality and capacity of the public telecommunication network.[2] These did little to destabilize the established telecommunication institutional structure until the 1960s in the United States and later in Europe. The widespread structural and organizational changes in the industry since then have been attributed to the diffusion of digital technologies and the application of microelectronics to telecommunication. But it was not until the early 1980s that most PTOs began to diffuse digital equipment throughout their national networks. Digital services are based on the conversion of information, voice, data, image and text into binary code which can be recognized by computers. Three main components of the telecommunication network have been affected by the digitalization of the telecommunication network: switching, transmission and terminal equipment.

Network digitalization has been regarded as a radical innovation following some fifty years after the electromechanical revolution in telecommunication. It has been argued by Freeman that the electromechanical paradigm was initiated by the invention of electromagnetism which enabled the reproduction of sound (telegraphy) or voice (telephony) at a distance. Similarly, the digital paradigm is based on computing techniques which convert information into the digital form required for high-speed processing and transmission (Freeman 1988). In this sense, the introduction of digital technologies has been treated as a radical, and perhaps even revolutionary, innovation. Nevertheless, the first use of digital technologies within the telecommunication network – namely Pulse Code Modulation – was originally envisaged simply as a quality enhancement to improve the capacity of overcrowded circuits.[3]

Schumpeter argued that 'Revolutions are not strictly incessant; they occur in discrete rushes which are separated from each other by spans of comparative quiet. The process as a whole works incessantly however, in the sense that there is always either a revolution or absorption of the results of revolution' (1943: 83). In this recent phase of innovation based on the digitalization of the telecommunication network, it is clear that Schumpeter's *creative gales of destruction* are putting strains on the existing institutional structure. However, analysis of the implications of the interdependency of technical and institutional change needs to focus on more than a single cluster of innovations such as digital technologies. The process of creative destruction refers to one in which 'innovation constantly reshapes the very market structures out of which it arises in an evolutionary way' (Scherer and Perlman 1992: 1). This continuous incremental process of technical change has had profound effects on the structure of the telecommunication industry throughout its history. Technical developments, in turn, have been shaped by the political and economic interests of the telecommunication equipment and service producers.

Gabel and Trebing, for example, have shown the ways in which the erosion of established market structures in telecommunication in the United States occurred during the period between the 1850s to the 1960s. Their analyses give careful consideration to the flexibility and responsiveness of public and private institutions which occurred in parallel with technical innovations. Trebing has argued that, in the United States, the first major challenge to the incumbent monopolist, AT&T, was evident in 1946 as a result of the developments in microwave technology. AT&T responded by attempting to integrate microwave capability within its nation-wide network and succeeded in delaying the competitive threat for almost a decade (Gabel 1969; Trebing 1969a). Microwave technologies were used to provide analogue, rather than digital, communication services at this time.

Adherents of the Idealist and the Strategic models of technical and institutional change agree that there are dynamic features in the telecommunication market which create instability. They disagree on the outcome of such instability. In the former model, organizations with substantial investment in earlier generations of technology, coupled with the inflexibility resulting from large-scale monopoly operation, see their command of the marketplace competed away. Hence, the Idealist model sees a market emerging from a competitive process in which only the most innovative and customer-responsive firms survive. The marketplace is regarded as one in which the strengths and weaknesses of the competitors are differentiated only by the technologies available at the time of entry, the relative efficiencies of their operations, and the strategic benefits that flow from alliances which compensate for changing technical and managerial competencies.

In the alternative Strategic model, the same process of competitive rivalry produces monopolistic firms and new market distortions. The incumbent monopolists appropriate technical innovations by creating barriers to entry, entering alliances and adopting a variety of other mechanisms to preserve their historical strength. The national PTOs, as incumbents, are not equally successful in their bid to retain or expand their market, and new dominant players can emerge as a result of strategic coalitions. The present transition in the telecommunication market is not from monopoly to competition, but from monopoly to strategic oligopoly.

In the 1970s and 1980s the firms that found themselves excluded from the traditional telecommunication club formed an alternative electronics coalition (Noam 1987). An unstable pattern of technical change was accompanied by commercial pressures to reorganize telecommunication and to expose the incumbents to competition. In the United States this resulted in the beginnings of erosion of the AT&T market power by operators of private networks. Constructed initially to gain the benefits of digital technologies before they were widely diffused throughout the public switched telecommunication network, private networks allowed large corporations to meet their communication requirements in-house by constructing their own networks. These companies used capacity leased from facilities-based carriers, and, in some cases, they built their own facilities. Despite these developments, neither the facilities-based carriers such as AT&T, nor the telecommunication equipment manufacturers have seen their market power eliminated completely. The impact of competition has been extremely strong in some segments of the service and equipment markets and extremely weak in others. The important issue for policy is whether the overall trend is towards the further erosion of the ability to monopolize markets or towards a new exclusive club with some old and new members – that is, towards strategic oligopoly.

In this context, the following questions need to be addressed. Is the intelligent network a strategy deployed by the traditional PTOs to protect their pre-eminent status? Can the intelligent network architecture be shaped successfully to meet this objective, or will it support the ambitions of the PTOs' competitors? In this chapter, the way in which the intelligent network was conceived as a design for the next generation of investment in the public telecommunication network is considered. The analysis provides a basis upon which to assess whether the intelligent network design is creating the possibility of providing flexible, low-cost services for *all* users. Alternatively, the new design may be the technical expression of a process of restructuring whereby the contours of global telecommunication networks are redefined mainly in the light of the multinational firms' requirements.

These questions raise issues concerning how primary design considerations and their associated costs are shaping the evolution of public telecommunication networks. Evidence in support of the predominance of designs associated with multinational firms' requirements need not demonstrate that all, or even the majority, of these firms' requirements are being met in a form that would satisfy their demand. It need only establish that these requirements are the predominant criteria that are shaping technical compromises within the innovation process.

The intelligent network has been described as a telecommunication network services control architecture (Ambrosch et al. 1989). Technical experts have argued that the initial aim in designing the new architecture was to breathe new life into the public switched telecommunication network. In the industrialized western markets, public switched telephone services appeared to be reaching saturation point and growth in real-time voice services had stabilized at a relatively low level. New advanced data, text and image services would be required to generate new revenues.

The public switched telecommunication network evolved throughout the world as a hierarchically structured, relatively inflexible system. Voice and data signals were routed up and down a hierarchy of local, trunk and transit switches – a configuration which continues to prevail throughout the United States and European countries. The challenge has been to transform this relatively rigid network into a flexible, intelligent telematics infrastructure. When the intelligent network is presented as a technical solution to problems of service control it appears to have the attributes of an innovation which is simply the latest in a series of engineering developments. However, its implications for the ways in which telecommunication networks evolve are far-reaching. In fact, this set of technical innovations and the institutional innovations that accompany them represent a turning point in network evolution.

## Debating the Technical Architecture

The generic intelligent network architecture was pioneered in the United States in 1985 by Bellcore (Bell Communications Research) at the request of Ameritech, one of the Regional Bell Operating Companies (RBOCs). Bellcore has been jointly owned by the RBOCs since they were created by the implementation of the AT&T divestiture agreement in 1984. The events which led to the structure of the US telecommunication market which had emerged at this time are recounted in Chapter 3. In February 1985, a request for information for a concept was submitted with the following objectives: to support the rapid introduction of new services in the network; to help establish equipment and interface standards to give the RBOCs the widest possible choice of products; and to create opportunities for non-RBOC providers to offer services that would stimulate network usage. The Feature Node concept originally promoted by the RBOCs was adopted by Bellcore and became the intelligent network concept (Ambrosch et al. 1989).

The stated aim was to increase the rapidity of new service introduction, encourage the development of competitive equipment and products, and to open the public network to greater use by non-RBOC service suppliers. The challenge was to create a feature-rich public network. This would be achieved by transforming the hierarchical system of switches into a flexible, complex infrastructure. The RBOCs wanted a network architecture with sufficient intelligence to recognize the origin, required processing and destination of calls. This would require considerably greater computerized data-processing capability than was embedded in the public switched telecommunication network which they inherited following the divestiture of AT&T in 1984.

Bellcore announced a conceptual model called Intelligent Network-1 (IN/1), and the term 'intelligent network' gained currency in the technical and trade literature. It became a generic term for networks characterized by distributed intelligence and the definition of interfaces and services in ways that departed from current practice (Magedanz and Popescu-Zeletin 1991).

Initially, the intelligent network was associated with freephone services, virtual private networks, credit card validation, universal access numbers, and call-back-when-free services in the United States. By 1987, a version of the intelligent network concept called Intelligent Network+2 (IN/2) had been introduced. In this architecture, services were to be provided completely independently of the physical structure of the network, a design which would require significant changes in existing switching capabilities. Bellcore subsequently announced an interim stage called Intelligent Network+1 (IN+1) to provide an initial way of meeting the RBOCs' specifications.[4] This three-stage implementation model is the same as that which has been under discussion for the introduction of *intelligent* mobile radio-based services – for example, Universal Personal Telecommunication (UPT) – while the Bellcore model was discussed at first mainly with regard to the fixed terrestrial telecommunication network.[5]

Bellcore's objectives for the intelligent network were described as falling into three categories: the development of a flexible network architecture, the implementation of standard network interfaces, and increased speed in advanced service introduction. The specifications which have emerged since 1985 represent alternative ways of meeting these objectives. Designed to permit open access to the network, the RBOCs claimed that the new architecture would help to increase their responsiveness to customers. Other service providers would also be able to use the public network to supply competing services, which would generate new traffic and revenues for the network owners.

The intelligent network has come to imply different telecommunication development strategies for network operators, equipment suppliers, service users and regulators in the United States and in Europe. Most agree that it refers to an environment where services can be provided independently of the physical structure of a network – that is, services are supported by logical networks and controlled by signalling of databases where the intelligence or service attributes reside. However, there is much uncertainty as to the modifications in the basic network architecture that are needed at the implementation stage and the implications for suppliers in the telecommunication and computing industries. There is even greater uncertainty as to the

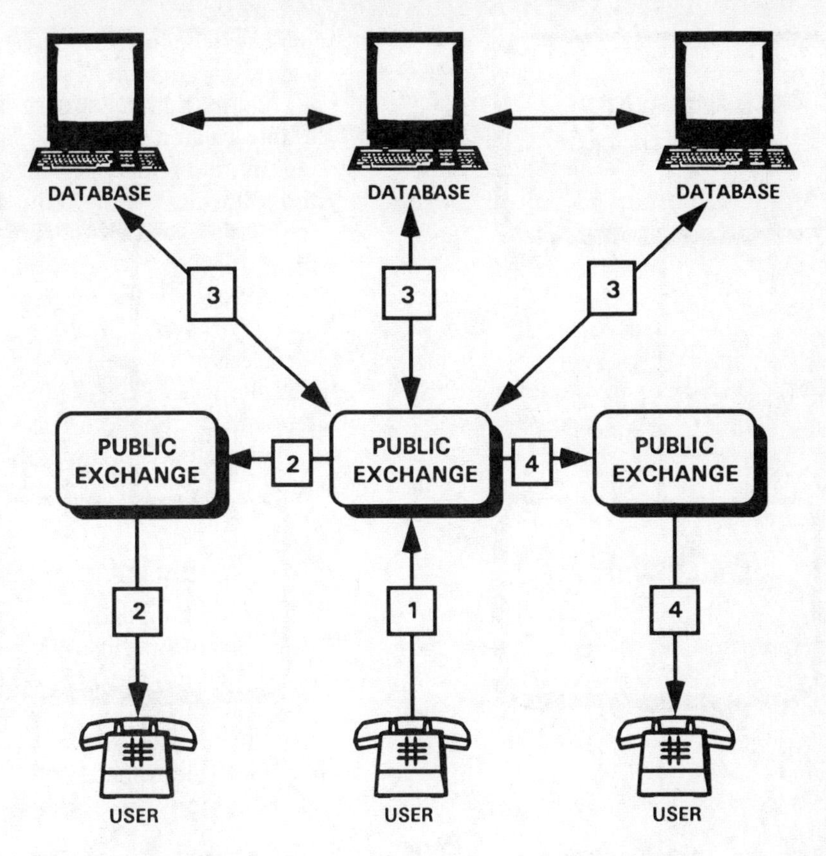


1. User makes call which is connected to the nearest public exchange.

2. If no intelligent network feature is required, the exchange routes the call to its final destination.

3. If an intelligent network feature is required, the exchange consults one of a series of databases about what to do. Messages are passed between the exchange and computerised database using a fast signalling system. The call itself is held at the exchange.

4. When the message is received from the database, the exchange routes the call to its final destination.

Figure 2.1 *The architecture of an intelligent network (*Financial Times, *1989c)*

changes telecommunication users are likely to see in the terms and conditions of access to public and private networks.

The International Telecommunication Union's Consultative Committee for International Telegraph and Telephone (CCITT), the organ charged

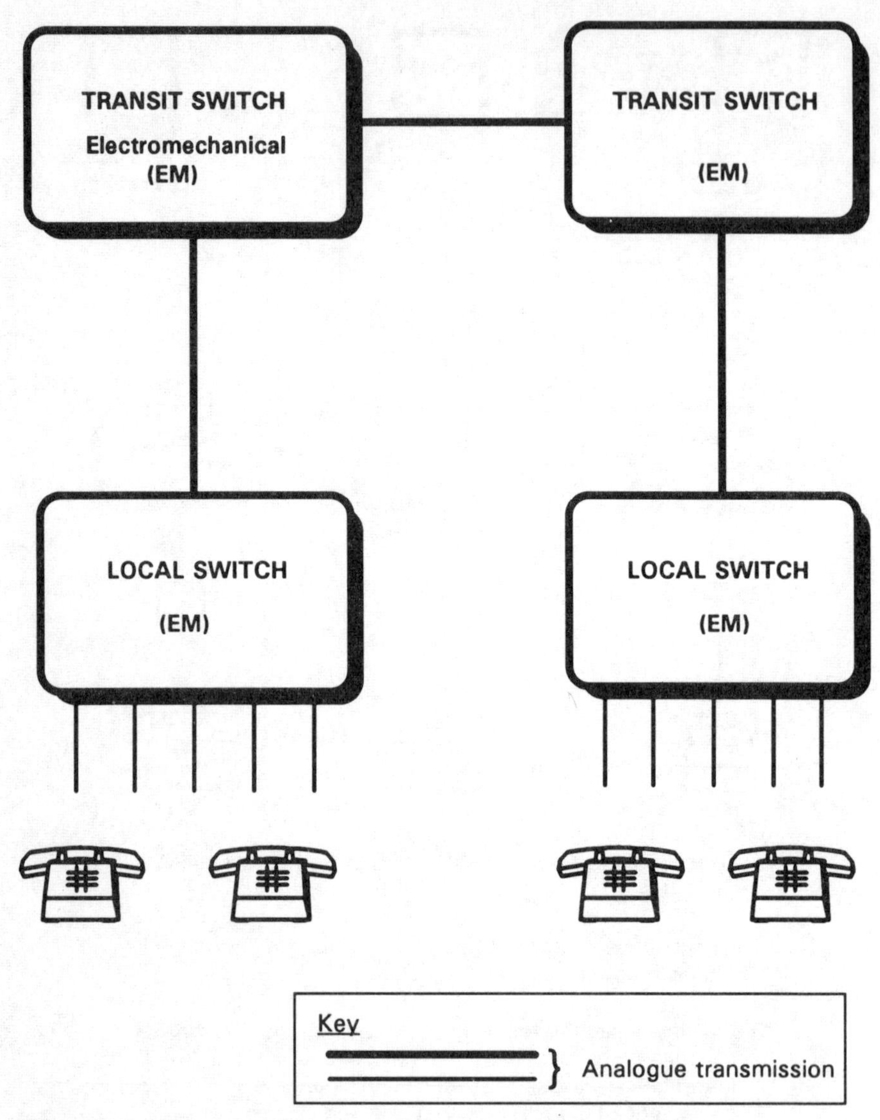


Figure 2.2 *The telecommunication network, 1920s–1970s*

with the development of international telecommunication standards,[6] defines the intelligent network as

> a new architectural concept for all telecommunications networks. Intelligent Network (IN) aims to ease the introduction of services based on more flexibility and new capabilities. Currently, the subject of IN is motivated by the interests of telecommunications services providers to rapidly, cost effectively and differentially satisfy their existing and potential market needs for services. Also, these service providers seek to improve the quality and reduce the cost of network/service operations and management. (Consultative Committee for International Telegraph and Telephone, International Telecommunication Union 1990: 4)

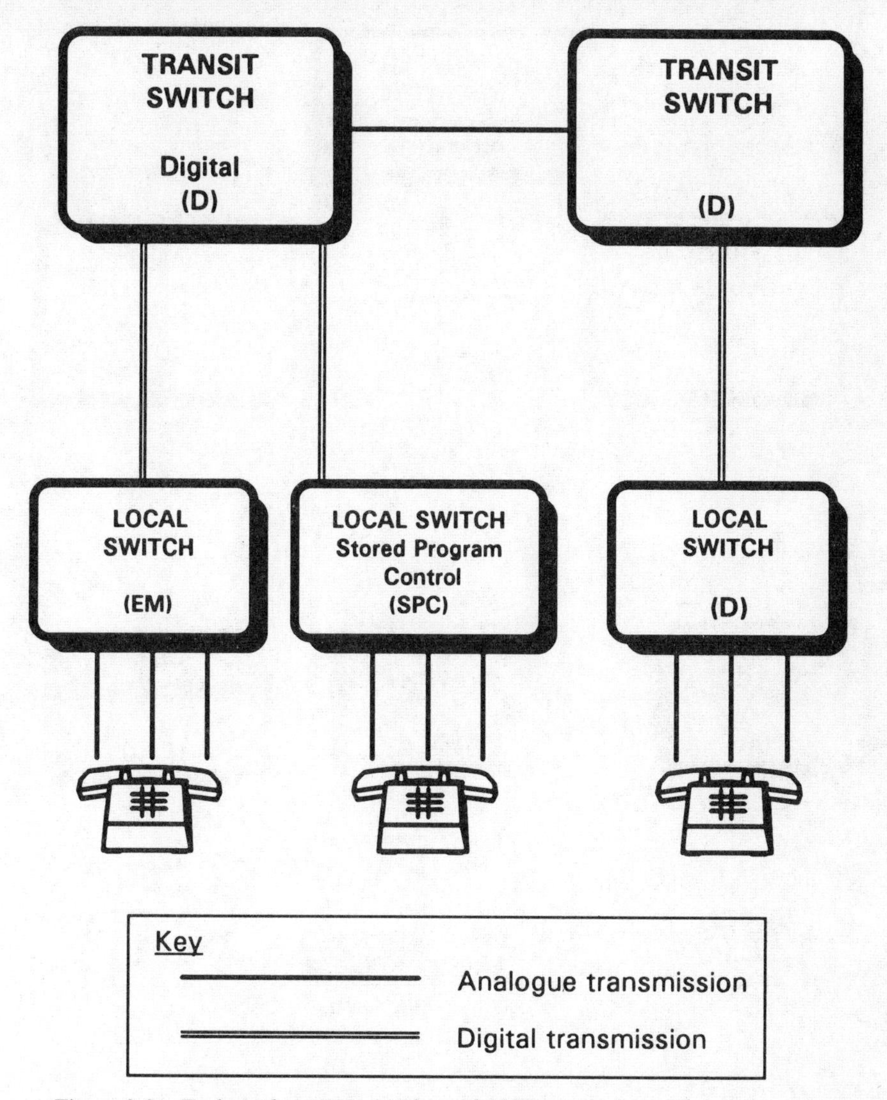


Figure 2.3 *Technical coexistence, the mid-1980s network*

The intelligent network architecture can be applied to any type of network. Figure 2.1 shows the main components of the new architecture in simplified form as they appeared in the press in 1989. The original design concept of the intelligent network was the introduction of databases into the relationship between the user (or competitive service supplier) and the public switch or exchange.

Figure 2.2 schematically depicts the hierarchical network which prevailed from the 1920s until well into the 1970s in both the United States and Europe. This network was comprised of a series of linked electromechanical switches and analogue transmission facilities.

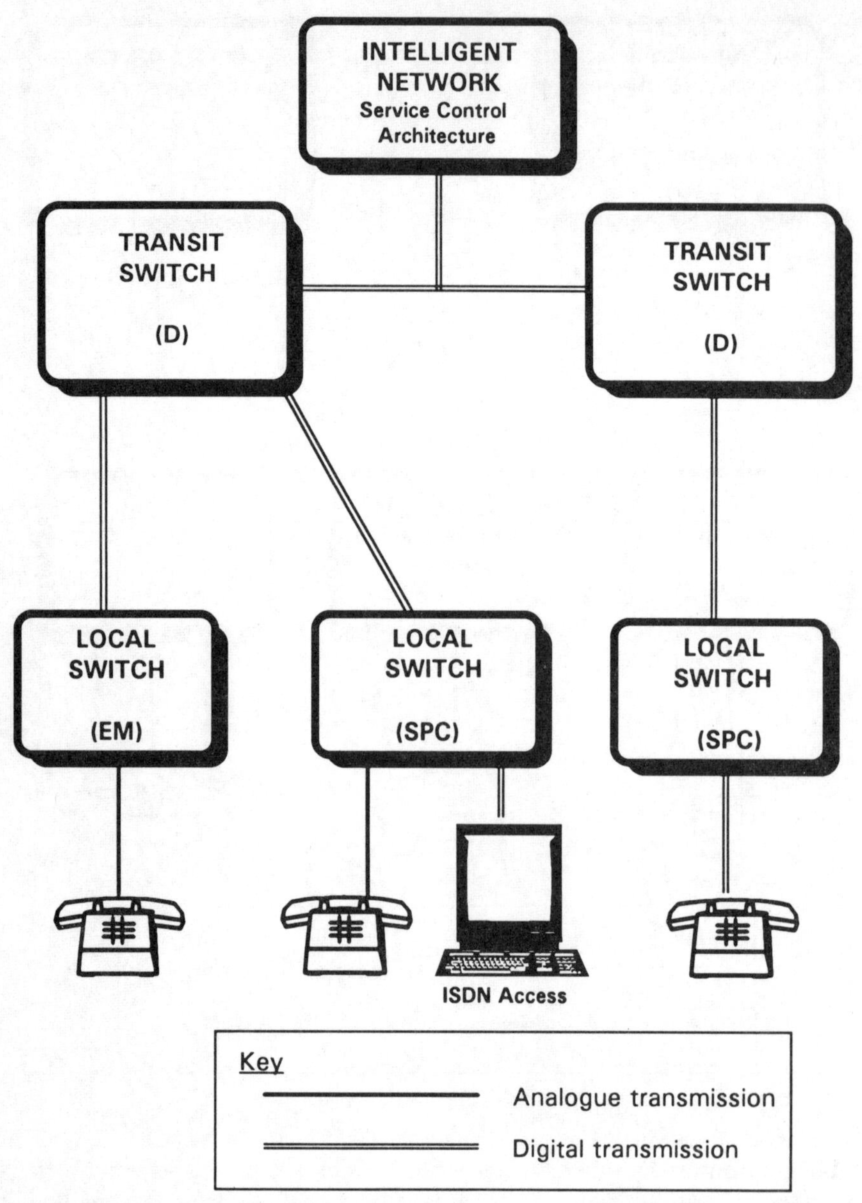


Figure 2.4 *The Service Control Architecture*

By the mid-1980s the public switched telecommunication network had begun to incorporate digital technologies and computerized switches. Figure 2.3 shows that a mix of analogue and digital links supporting transmission had come into use and that electromechanical (EM), computerized Stored Program Control (SPC) and Digital Switching (D) had begun to coexist.

The generic intelligent network architecture involves more than an all-digital infrastructure. It requires the introduction of a Service Control Architecture (see Figure 2.4) which includes new functional elements.[7] The design, control, cost and combination of these elements are critical to the resources available to the PTOs to respond to, and control, new entry into their markets.

Figure 2.5 shows the main functional elements or components of the generic intelligent network. The Service Management System (SMS) runs on a computer. The SMS updates Service Control Points (SCPs) with new data or application programs and collects statistics about the performance of the network. In principle, access to the SMS would let a customer directly control the service attributes via a terminal.

The SCPs come into play to differentiate between basic and newer supplementary services when the latter are introduced into the network and to ensure that signals are interpreted and routed in the appropriate way. The service programs and data are updated by the SMS. The SCP can be either a modified public telecommunication switch or a commercial computer which is connected to the main network. In either case, the critical factor is reliable and efficient access to the databases.

The Signal Transfer Point (STP) shown in Figure 2.5 is part of the Common Channel Signalling System No. 7 (CCSS7) network, a standardized communication interface which links the network's intelligence to the switches embedded in the rest of the physical network.[8] This element can be integrated with the switch or provided on a stand-alone basis. The standardized protocol for CCSS7 includes a format for messages flowing within the network.[9] These trigger its various components into action. The Service Switching Point (SSP) is the point of network access for all service users as well as competitive service providers.

Common channel signalling is a method of providing control in a telecommunication network. This form of signalling reduces the cost required for separate signalling units in each switch and permits greater flexibility for the development of future services. Signalling System No. 7 refers to the CCITT recommendation for common channel signalling which is designed for use in networks conforming to Integrated Services Digital Network (ISDN) standards, but it is not limited to this configuration. This signalling system provides internal control and network intelligence but it does not transfer data. The recommendation was developed in the 1970s and it has been used to define the format and content of packets of data on the 'D Channel' in the ISDN configuration.[10] The ISDN recommendations were approved by the CCITT in 1984 and revised in 1988. The main objectives of ISDN were to transform analogue public telecommunication networks into digital networks, to provide voice telephone network users with an opportunity to access a range of advanced services, and to create a public telecommunication infrastructure that could provide a focal point for further innovations in information and communication technologies.[11] ISDN requires the installation of digital switching, and the format of service

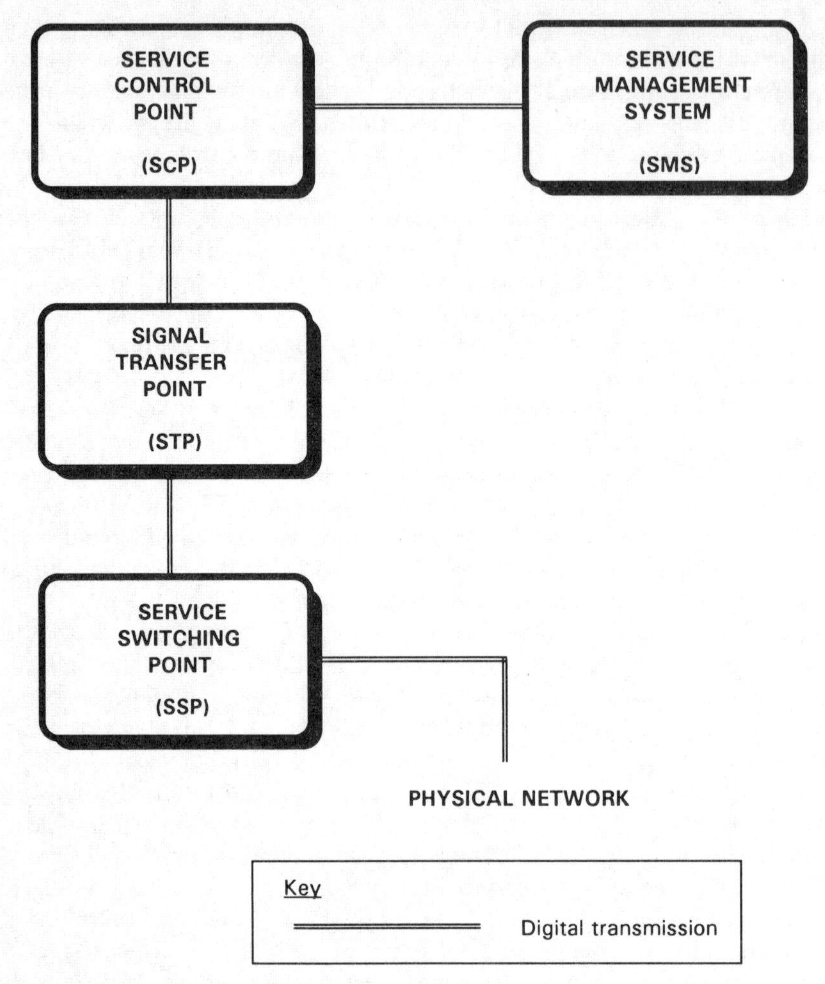


Figure 2.5 *Intelligent network components*

control is specified by CCSS7. In the ISDN configuration, user data travel on 'B Channels' at 64 kbit/s. CCSS7 is often referred to as out-of-band signalling, a characteristic which is fundamental to the provision of both ISDN and intelligent networks. This method of signalling allows network control signals to be conveyed outside the frequency band which is normally used for user data or message transmission.

This is the technical innovation which creates flexibility in the provision of services. Darmaros argues that ISDN innovation has been an example of a service configuration that is strongly supply-dominated. The new technology is being introduced in a characteristically technology-push manner, with little regard to economic reality. In addition, 'the correcting ex-post effects of markets . . . are very difficult to realise, especially as long as traditional monopolists remain the dominant players' (Darmaros 1992: 461,

464). In some countries, ISDN has been used by the PTOs as a marketing device to justify the costly introduction of CCSS7 within their networks. In other countries this has not been the case, and the introduction of advanced signalling has been justified on the basis of the need for a flexible way of controlling complex services such as those planned for the intelligent network environment. The ISDN standards specify a relatively inflexible network for the transmission of voice and data signals,[12] and this has been criticized by many large telecommunication users who wish to access considerably greater bandwidth than that offered by the ISDN configuration.

The history of ISDN is not the central focus of this book. However, there is a sense in which the intelligent network architecture represents a renewed attempt to alleviate the constraints on flexibility in the configuration and use of networks created by the earlier standardization of ISDN. The intelligent network provides one of many possible solutions to the rigidities of an earlier network design configuration. It proposes to achieve this by integrating the use of advanced out-of-band signalling with a variety of possible bandwidths, network control features and service applications. The main components of the intelligent network can be configured in different ways. These differences affect where intelligence or processing power is located throughout the physical network; for example, modifications to local, trunk and transit switches or centralization of intelligence in a small number of network nodes.

The functions performed by public switches in the conventional network can, in principle, be unbundled in an intelligent network configuration. For example, voice traffic can be handled by the Service Switching Points. When more advanced services are involved the Service Control Points can be invoked. In the intelligent network there are two kinds of traffic flows: the service traffic, and the information exchanged between the network nodes which carry orders to the Service Switching Point for the routing and processing of traffic.[13]

If service information is centralized in the Service Control Point, the public network might gain in flexibility. The need to upgrade the software in existing switches to implement a new service could be avoided. In some intelligent network configurations, only one or two Service Control Points need to be upgraded to support a new service. Using CCSS7 links to access information, which is then dispersed throughout the network on request, new services can be introduced rapidly, matching the availability previously achieved only by long-established services. This is the technical argument in support of the benefits of the generic intelligent network. In principle, there should be substantial savings for all customers, earlier introduction of advanced services, and increased openness of the public network to competing service suppliers.

Control of the Service Management System (SMS) lies at the heart of the issue as to whether the implementation of the intelligent network will create opportunities for the restructuring of the telecommunication market in a way that resembles the expectations of the Idealist or the Strategic model. For example, if the network can be accessed and used by challengers to the PTOs, what technical or economic threats will this create for the incumbents in the market? How will the PTOs recover investment in the computerized systems which allow advanced services to be introduced?

## Network Design and Market Boundary Definitions

Arguments as to how the boundaries which define viable market territories for the PTOs and their competitors will change, how the intelligent network should be introduced and by whom, generally take place in the sanctity of the fora charged with establishing technical standards for the public telecommunication network. For example, in December 1992 the Department of Trade and Industry in the United Kingdom commented that

> decisions taken by both the international standards fora, on defining Intelligent Network architectures, and by the Public Telecommunication Operators, in evolving their current networks towards this architecture, will have an impact on the technical complexity of the access to which service providers may wish to interconnect. . . . There has so far been minimum consideration on the need to support access for competitor providers within the standards fora, mainly because the representation from this sector is virtually non existent. This is hardly surprising since we are dealing with a competitive activity that *largely does not exist and as yet is relatively undefined*. (Department of Trade and Industry 1992: 2, 8; emphasis added)

Comments on the definition of network interface standards and protocols provide a flavour of the political and economic interests at stake since it is these standards that could enable open access to the public network. Documents prepared by the CCITT working party charged with development of an internationally agreed standards document provide a good illustration.

> An essential attribute for IN [intelligent network] evolution is the modularization of functionality and information, thereby providing flexibility in its distribution while maintaining *unified control* giving a single integrated vision. (Consultative Committee for International Telegraph and Telephone, International Telecommunication Union 1990: 5; emphasis added)

The reference to a unified control structure is reminiscent of historical justifications offered by PTOs whenever their monopoly status has been challenged (Babe 1990; Davies 1991; Mansell et al. 1990). For example, in the United States AT&T argued after 1885 that unified ownership of telephone systems would help to guarantee the quality and standardized design of telecommunication equipment (Garnett 1985). It is evident that standards for the intelligent network could be designed to favour a competitive multi-vendor environment. They could also be defined in such a way that they provide the PTOs with a new tool to maintain control over access to the public telecommunication infrastructure. This could strengthen their ability to plan the evolution of the network, but it could also erode their market share if new entrants gain access to the intelligent capacity of the network and use this to develop competing services.

As conceived by the CCITT, the intelligent network architecture is applicable to mobile, packet switched public data networks and ISDNs. As a result, it does not apply only to the fixed terrestrial infrastructure that was designed originally to support voice services. The CCITT has suggested that the transformation to the intelligent network will be a long-term process in which 'the specification and deployment of networks that meet *all* of the objectives and comply *fully* with the definition of the target IN [intelligent network which] will take many years and will be the subject of a long term architecture study' (Consultative Committee for International Telegraph and Telephone, International Telecommunication Union 1990: 5; emphasis added).

Initial CCITT standardization work focused on the development of an architecture for distributed intelligence under unified network control. This work is to be followed by standards for service logic and parameters that can be controlled by the PTOs as the primary service providers. Only then will the last phase of standardization become concerned with the open-ended features of the intelligent network (Consultative Committee for International Telegraph and Telephone, International Telecommunication Union 1990).

The intelligent capabilities of networks could be located in a private network operator's equipment at the periphery of the network. Furthermore, as liberalization of the use of public telecommunication networks increases, opportunities for mingling voice and data traffic in private and public networks have expanded. New network designs partly reflect the pressures upon PTOs to make their networks more open and flexible to competing service providers (Mansell and Hawkins 1992). For example, third-party service vendors and users need access to some components of the intelligent network if they are to control programming of the databases and fill customer orders for communication and information-processing services. But investment by the PTOs in an open network could create a basis for competing service providers to appropriate the added value accruing from the public network operator's initial investment.

The PTOs and their suppliers are extremely unlikely simply to relinquish their present dominance in the marketplace. The malleable design of the intelligent network architecture offers one way to retain control. The network suppliers, their competitors and the end-users can influence the process of network design to the extent that they are able to influence decisions with respect to technical standards. However, there is considerable evidence that the impact of most users is negligible in standards setting fora (Hawkins 1992a).

The manufacturers of switching equipment for public telecommunication networks also have a strong interest in protecting their markets. But the spin-off effects of the intelligent network for equipment manufacturers in the computing industry could be substantial. For firms in the computing and software services industry, the intelligent network is a way of encouraging a multi-vendor environment and creating opportunities for increased sales of hardware and software needed to complete the transition to a fully intelligent public (or private) network infrastructure.

In the mid-1980s there was considerable optimism on the part of industry observers that differences between the domains of computing and telecommunication would be resolved as a result of the forces of technical convergence between previously relatively distinct engineering competencies. Discussions in voluntary supplier and user industry fora were expected to provide the venue for convergence and the shifts in the boundaries of industry participation in telecommunication that would result. By the early 1990s, however, the role of computer and software vendors was less certain. An interview with a representative of IBM suggested that its participation in the intelligent network market was not as secure as it had thought when technical convergence seemed inevitable some five years earlier.

Although telecommunication is becoming a global enterprise, the components of public networks continue to be designed and implemented at the national level. Differences in design, interface standards and the unbundling of intelligent network functionality occur at this level. Hence, the test of the Idealist and the Strategic models comes from an understanding of the design and implementation of the intelligent network by PTOs and equipment manufacturers in the national (or regional) context. The international standards-setting organizations create the broad contours of the opportunities for market restructuring and, in this sense, they bias market developments through their standards-setting activities. However, these can undergo substantial modification at the regional or national level, and equipment manufacturers and PTOs frequently move to implement innovative network designs long before the standards-setting process at the international, regional or national level has been completed.

For some of the large telecommunication equipment manufacturers, the intelligent network represents a path toward a growing market provided they are able to maintain control over the design and implementation process. In principle, the design of the physical parts of the public telecommunication network could be completely separated from the creation of telecommunication services. This view of the design challenge is consistent with the early version of the generic Bellcore intelligent network. The switch manufacturers and PTOs must decide how and when to introduce products – for example, digital exchanges, Signal Transfer Points and so on. These are difficult choices when the design of hardware and software is undergoing constant modification and when new investment in intelligent equipment may jeopardize the ability to control network access.

Firms from the computer industry tend to see certain aspects of the intelligent network such as the Service Management System (see Figure 2.5) as being closely related to their core business. But the PTOs have not turned to computer firms automatically to supply Service Management Systems even though they are based on powerful computers which hold the database(s). The intelligent network components are costly and complex. For example, the performance standard set for Service Control Point down time must not exceed three minutes per year. The Service Management System is also an essential network element and it is the critical arbitrator of control over the network. As a result, this element indirectly controls the direction and speed of restructuring in the telecommunication equipment and service markets.

The implementation of the generic model of the intelligent network would have broken up the public network into interchangeable parts. These could have been offered on the market as separate commodities. These 'commodities' could then have become the basis for an unbundled and extremely open public telecommunication infrastructure that, from a technical point of view, would meet the criteria required by the Idealist model. The network 'commodities' could be re-packaged by the PTOs and their competitors to stimulate greater competition. This scenario was in line with the prevailing regulatory trend in the United States, which called for the implementation of Open Network Architectures (ONA) and is discussed more fully in Chapter 3.

From Bellcore's perspective in the mid-1980s, the commoditization of network components would extend its strategy of building partnerships across the information sector. Between 1989 and 1990 Bellcore partnerships included ventures with computer, video, microelectronics, switch manufacturers and a host of related information-technology developers.[14] Information sharing and external procurement were becoming part of Bellcore's business to be responsive to the RBOCs' goals of breaking out of their telecommunication *bit-transporter* role. A bit-transporter would be responsible for transmitting messages in the form of voice, data or image signals through the network, leaving the added value generated by the processing of such messages to be captured by other service providers who would use the network as a form of 'common carriage'.[15]

From the point of view of Bellcore's technology designers, the RBOCs would want to pick and choose among equipment suppliers. The original generic intelligent network would allow them to achieve this because open standardized interfaces would be agreed for different components of the network. Competing service suppliers could use these components to build their own services, and customers; for example, private network operators, in a strongly competitive market, would be able to opt for management of their networks by the PTOs or third-party operators based on their distinctive product offerings and efficiencies of supply.

Figure 2.6 shows that there are several possible relationships between suppliers that could enter the intelligent network marketplace to offer complete service solutions to the customer. But the way in which these suppliers might rank the commercial viability of the alternatives would have a major impact on the eventual design of services together with their accessibility and cost.

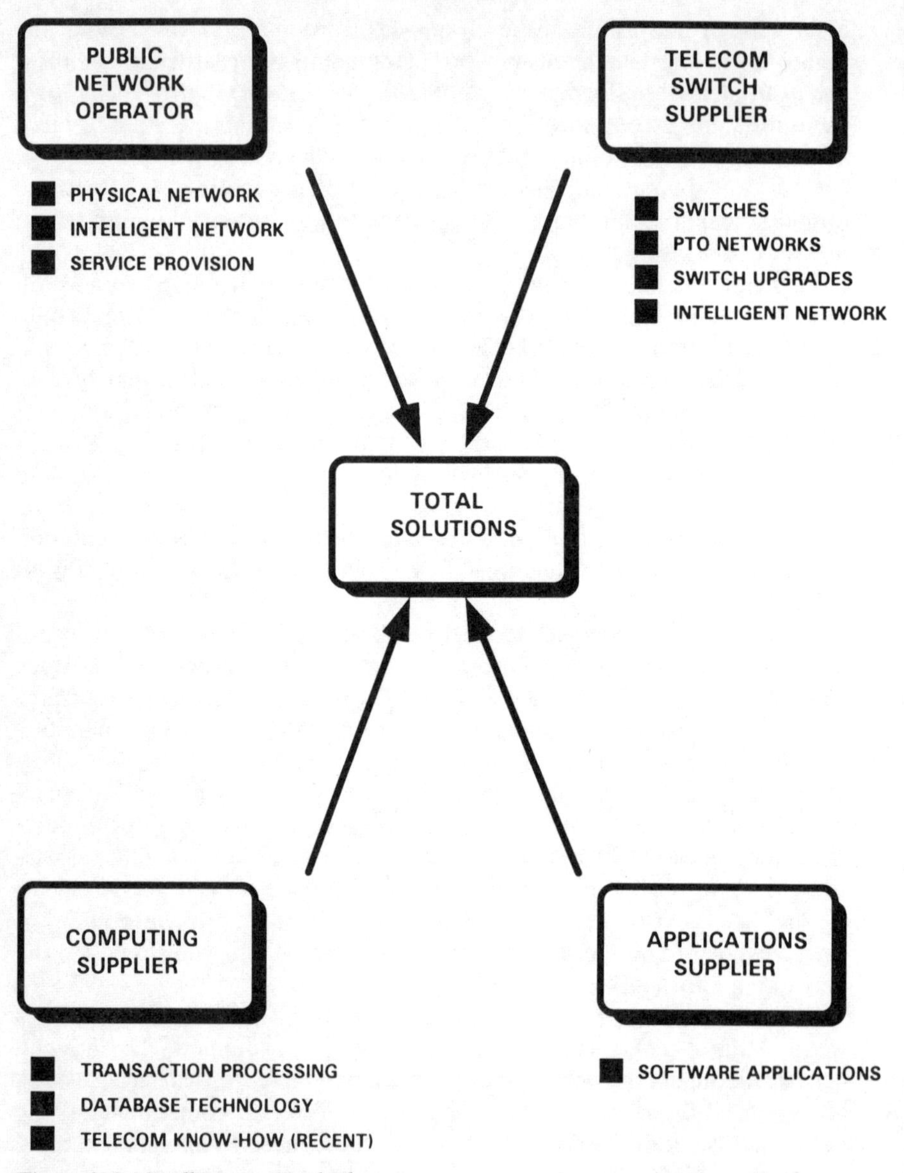


Figure 2.6 *Intelligent network players*

The feasibility of substitution between the intelligence residing in peripheral terminal equipment controlled by the user and/or the private networks of competing service providers and the public switched telecommunication network is an important factor that shapes the dominant trends in public network modernization (Konig 1989). The intelligent network is supported by software which many PTOs do not have the competence to develop or maintain. The intelligent network generic model would mean, theoretically, that public telecommunication switches would become less dedicated to performing specific software-driven functions. As a result, general purpose computers could be used instead of, or in tandem with, existing and upgraded public telecommunication switches.

There are two extreme scenarios for the intelligent network design. On the one hand, the public network could lose all of its intelligence. On the other, it could retain some portion of the software functionality in the switches. The advantages and disadvantages of these alternatives depend upon the allocation of costs to different functions as well as upon the respective competencies in software development of the PTOs and the equipment manufacturers.

If functionality is unbundled within the network, then decisions must be taken as to how to allocate the costs of network components. The commercially significant question that arises in this scenario is what share of these costs should be recovered by revenues generated by PTO supplied services and from firms purchasing network resources to launch competing services? This question is fundamental to which network operators and suppliers will command control over network access. Because of the joint and common cost characteristics of network supply the answer to this question is not straightforward. As Trebing observed at the end of the 1960s during an earlier phase of market restructuring in the United States,

> the combined effect of these cost characteristics is to introduce economic forces which tend both to limit the number of firms in a given market and to increase the basis for discriminatory pricing. Discretion in the assignment of costs does not necessarily mean that reasonable assignments which relate plant and facility usage to particular classes or types of customers cannot be made. Rather, the strategic question is, who shall make these judgments regarding cost assignment . . . clearly, the public interest would dictate that the duly constituted administrative agency should exercise this responsibility. (Trebing 1969a: 300)

Although technical change has created opportunities for new entry in the telecommunication market beyond the scale envisaged even in the United States at the end of the 1960s, it has not eradicated opportunities to achieve network closure through standards setting, pricing policy, and network access and interconnect conditions. These issues are taken up in greater detail in Chapters 10 and 11.

In 1987 the trade press reported that the design of the intelligent network would offer an evolutionary path towards open networks in the United States and in Europe (Gilhooly 1987). By 1991 the likelihood of this outcome was much less clear: 'Major carriers on both sides of the Atlantic are ploughing ahead with the delivery of key IN [intelligent network] services, such as freephone and virtual private networking, using *proprietary* platforms to do so . . . the standardised intelligent network is now raising as many *fears* as hopes' (Finnie 1991; emphasis added). Major design issues would need to be resolved if the intelligent network were to provide the basis for a loosening of the PTOs' dominant position in the telecommunication supply industry. These would include the weakening of the integrated vision of the public telecommunication network espoused by the CCITT, regional, and most of the European national telecommunication standards-setting fora, and a strengthening of the multi-vendor environment advocated by the computing industry and the service supply industry.

The factors affecting the negotiations that would resolve these issues are expressed primarily in technical terms but they embody a range of economic and political factors. A framework for investigating the tensions in the negotiation of the design of the intelligent network is presented in the following section. This framework focuses on the design parameters which will strengthen or weaken control by the still dominant PTOs over access to the electronic communication environment of the 1990s.

## Analytical Framework – Contending Network Designs

An investigation of the political economy of intelligent network design requires an analytical framework that links technical and non-technical factors. The framework must point to areas in which technical design choices become enmeshed with political and economic factors which influence the restructuring of institutions, including markets, firms and regulatory organizations.[16] It must also incorporate policy objectives.

In the European context, these are assumed to be improved efficiency in equipment manufacture and service supply and the extension and modernization of the public telecommunication infrastructure. These objectives are variously interpreted and weighted in the policies developed by the European Commission and the member states of the European Community. They are linked closely with alternative views as to how an industrial policy for the Single European Market (and the European Free Trade Area) can stimulate the competitiveness of the information technology and telecommunication industry. For telecommunication in Europe, the main public policy objectives are:

> to provide *users* with the multiple advanced services on the best cost terms. These services would combine information and communication technologies to give a potential explosion of possibilities, and so contribute to supporting the competitiveness of the business community; to put at the disposal of the European telecommunication *industry* a large market capable of providing it with the economies of scale required by the high cost of developing the new equipment, and the need to confront relentless world competition; to allow the *network operators* to address the demand for advanced telecommunication services, to encourage them to make the decisions to invest, and to accept the corresponding risks in doing so. (Carpentier et al. 1992: 90)

In this study, the framework also must provide a basis upon which the realism of the contending Idealist and Strategic models can be assessed.

### *The Political Economy of Network Design*

The endogenous economic growth model proposed by neo-classical economists investigates the behaviour and performance of stylized homogeneous firms to explain the determinants of competitiveness in a marketplace in which no single firm or organization is capable of undue control or influence over productive decisions. In contrast, an institutionalist perspective seeks to understand the dynamics of technical and institutional change under very imperfect market conditions where there are substantial variations in market power.[17] An emphasis on institutionalized market power is found in the political economy tradition in the communication field. For example, Williams has argued that: 'the major modern communication systems are now so evidently key institutions in advanced capitalist societies that they require the same kind of attention, at least initially, that is given to the institutions of industrial production and distribution' (1977: 136). The biases and imperfections of market conditions become embedded in institutions. These are expressed as power relations which operate in the design, selection and implementation of alternative technical systems. The production of products and services within the communication environment is the concrete manifestation of the dynamic processes at work in the generation and implementation of the public telecommunication infrastructure. In this sense, the telecommunication system is much more than a technical system composed of hardware and software, transmission links and switching apparatus. It is a technological system which embraces the technical artefacts and the institutions which shape its development, diffusion and use.[18]

The aim of a political economy of network design is to expose aspects of telecommunication network evolution that bias its development and become part of a complex system of institutionalized power relations.[19] In this study, the main focus is on the ways in which key private and public sector institutions – that is, telecommunication service suppliers, their equipment manufacturers, the major multinational telecommunication users, and the policy and regulatory apparatus – are shaping the design of the telecommunication infrastructure.[20] The political economy of the telecommunication network that is developed here offers an analysis which is complementary to related approaches to the relations of power and control that are embodied within the electronic communication environment. For example, equally important are approaches which focus directly on the way in which network evolution influences calling patterns, privacy of customer information, and the centralization or decentralization of control over information conveyed, or generated, by public network transactions (Samarajiva and Shields 1992).

Within any given technological paradigm, the cumulative character of the innovation process is shaped by the characteristics of institutions in the public and private spheres.[21] Successive iterations of the technical selection process within a given paradigm provide the mechanism for dynamic changes in supply and demand, and they structure the technical and institutional trajectories that emerge. As Dosi and Orsenigo argue, 'in contemporary mixed economies, market competition and other forms of more discretionary selection (such as choices by governments, financial institutions, etc.) sort out the behaviours, products, techniques, and organisational forms which – on some economic and/or institutional criteria – are "preferred"' (1988: 13). By conceiving of the innovation process in telecommunication network design as an instance of technical selection, analysis can focus upon 'the *relationships* between the various actors involved in a process in which all component phases are interlinked' (Gaffard 1992; emphasis added). The characteristics of internal and external relations among firms and other organizations are crucial determinants of innovative capacity and of the structuring of market power within the productive system. Although many have pointed to the need to investigate institutional factors which shape the technology selection process,[22] there has been virtually no work which focuses specifically on telecommunication network design issues.[23]

The analytical framework used in this study indicates parameters that are essential to the design and implementation of the intelligent network (Mansell and Jenkins 1992). This involves the specification of a limited number of factors which shape the emergence of dominant network designs. Metcalfe and Reeve argue that a framework can be developed by identifying 'patterns of complementarity and substitutability' between different technical alternatives (1990: 14). Like other technical systems, a telecommunication network can be analysed in terms of several different dimensions, for example, 'the artefacts in which it is embodied; the knowledge base which allows these artefacts to be produced; and the set of skills and competences which translate knowledge into artefact' (Metcalfe and Reeve 1990: 14).

Metcalfe and Boden see 'design configurations as the fundamental units around which to organise the discussion of technological change' (1992: 62).[24] The design configuration of an artefact can be used to shed light on the competitive strategies of firms and the constellation of knowledge and skills that is needed for production.

Another dimension can be added by building upon Braudel's observation that the effective exploitation of networks of ocean navigation, canals and roads depends on the existence of a supporting institutional infrastructure (Braudel 1984: 24). Braudel argued that the pace and ease of movement within any network system are conditioned, not by the maximum potential of the parts of a system, but by the minimum characteristics of its elements. He emphasized the structural and organizational features of technical networks. In order to examine the implications of the design of the intelligent network it is useful to consider the minimum conditions needed to achieve open access and use of the telecommunication network.

The framework provides a basis upon which to assess the Idealist and Strategic models in terms of their correspondence to the actual process of network design. In this study, design choices with regard to network interface standards, the unbundling of network intelligence, product differentiation, service competition, network access and control are the main features that are considered. Metcalfe and Reeve (1990) suggest that analysis of technical change should focus on key units of *productive transformation*. These could include, for example, network designs that alter the physical form of networks and their geographical location and reach. The essential transformation units should point to the determinants of boundaries between networks of productive activities. One such transformation unit is the minimum conditions required to develop the permeable transparent network envisaged by the Idealist model of telecommunication development. These are the open network interface standards.[25]

The generic model of the intelligent network initially assumed that the interface standards required to access the public telecommunication network would be put in place. Open interface standards could enable the PTOs to choose the technical components of the intelligent network from a variety of hardware and software vendors. Open interface standards could facilitate the interconnection of services provided by different PTOs or their competitors. Such standards also could be used to facilitate access by competing service suppliers to the unbundled intelligent functionality within the public telecommunication network.[26] But a report for the European Commission claims that 'the opening of the network resource control interface between switches and information providers is believed to be dangerous for network integrity and would make it possible for a third party to monitor, and potentially market, the data in the network' (European Telecommunication Consultancy Organization 1990: 47–8). This argument creates pressures for closed network interfaces and points to the reality of the Strategic model.

The framework used in this study directs attention to the economic and political factors which are linked to pressures to design open or closed interfaces. These interfaces are referred to as proprietary interfaces to indicate that they have been developed as a means of appropriating the tangible features of technical designs by organizations who may then seek to license their use. In some cases, such interface standards may become the dominant standard, in which case they may be more open than publicly agreed standards which are developed in national, regional or international fora. In the present study, the critical issue is the extent to which proprietary standards are used to restrict access to features of the intelligent network and, thereby, to strengthen the degree to which firms are able to appropriate control over technical innovations and so maintain, or expand, their position in the market.

The public- and private-sector organizations involved in the specification of interface standards do have technical reasons for their design decisions. Although 'design' is not the term commonly used by standardizers to describe their activities, most acknowledge that this is the activity in which they are engaged. Furthermore, there is a wide range of social, cultural, political and economic factors that inform these activities (Hawkins 1992b). Commercial decisions are taken before and during the standardization process. These influence the form and content of standards and, in turn, the unbundling of network intelligence, product differentiation and service competition strategies, network access conditions and, ultimately, the openness or closure of the intelligent network.

A political economy of standards setting which links institutional change with political agenda, government–industry relationships, and firm strategies has recently begun to develop (Hawkins 1992b; Reddy 1990). This approach has shown how proprietary and consensus standards are critical to complex systemic technologies (David and Greenstein 1990; Rosario and Schmidt 1991). This research has emphasized the important role of the standards-setting process, its institutions and participants. A further step is necessary if a political economy of telecommunication network evolution is to be developed in ways that link the determinants of technical and institutional change. This step is the incorporation of factors contributing to standards decisions within a wider analysis of network design and implementation. This is the challenge in the present study of the intelligent network.

### *Unravelling Network Designs and Visions*

The PTOs, in both the United States and Europe, have not moved as rapidly as initially expected to take advantage of the benefits promised by the generic intelligent network model. Instead, they have begun a process of partial and proprietary implementation of designs that differs considerably from the original model. Are their strategies explained by unresolved technical problems, or is the intelligent network threatening to challenge the existing distribution of market power and public network control? The acceptance or rejection of a network design is a reflection of an intricate mixture of technical and non-technical considerations. Like all complex and evolving systems there are unlikely to be straightforward linkages between design activities, standardization, the implementation of operational systems, and economic and political consequences. Nevertheless, a framework focusing on the conditions for open network access can be used to shed light on the use of technical design as a strategic tool of oligopolistic rivalry. The resulting analysis can contribute to policy questions such as whether emerging network designs exacerbate, or alleviate, unequal conditions of access to public telecommunication networks and whether new forms of public policy intervention should be considered. The key analytical question is 'From what sort of social relations does this artefact arise?' (Smythe 1972: 13).

The unravelling of the factors which are shaping the electronic communication environment of the 1990s presents a considerable challenge, and the analysis in this study is limited in several respects. It does not offer an in-depth review of all the competencies and strategies of the public and private organizations with a stake in the intelligent network. The comparative part of the study is limited to the main companies and policy-makers in the United Kingdom, France, Germany and Sweden. Developments in the United States are used to provide a context for the analysis of network design choices which have been taken in Europe.

The analysis is also limited by the time frame under investigation. The period 1988 to 1991 was one of widespread change in the structure and design of the public telecommunication network. However, the origins of these changes reach back to much earlier periods in the development of public networks in both the United States and Europe. Some of the constraints created by earlier strategies are discussed, but, in the following chapters, it will be important for the reader to recall that the design of the intelligent network has been shaped by three important historical factors.

The first is the pattern of investment in public networks. Intelligent network implementation strategies are shaped by the installed equipment base and the rate at which public operators move to replace this equipment or to upgrade it. For example, the penetration of advanced signalling systems, such as CCSS7, and the introduction of the ISDN, are crucial issues. The rate of penetration of digital connections and CCSS7 has differed in the United States, the United Kingdom, France and western Germany.[27] In networks without extensive CCSS7 implementation, the costs of migration to the intelligent network are greater than when this system with its out-of-band signalling capability has already been installed.

The availability of CCSS7 also determines the types of services that can be introduced.[28] Many advanced services can be supported either by the ISDN or the intelligent network configuration. The achievement of service independence from the physical structure of the network is linked to the availability of CCSS7. Standards define the interfaces which allow access to the nodes of the network and they provide the means for 'access to and displacement of the intelligence of the network' (Berben 1988: 3).

The second historical factor relates to the technical complexity of network design and operation. This is an area where PTOs and manufacturers can *manage* access to the public network. Pre-intelligent network management systems have not been designed to cope with full equal access by competing service vendors. Open network access requires investment by PTOs, and this has a bearing on the timing of new investment in intelligent network features. The PTOs have argued historically that public networks must be designed to handle traffic under the control of a central administration.[29] If traffic loading and network control are decentralized, they suggest that the reliability of the network could be threatened.[30] This argument has been used whenever the PTOs' monopoly has been challenged, whether in the United States or in Europe.

Some engineers believe that unexpected bursts of traffic in an intelligent network environment could grind the network to a halt. Others are not convinced that the software is available to ensure network operation in real-time. Unlike computing services, the telecommunication network cannot be taken out of service to correct faults. Although the validity of these technical defences of unified network control has been challenged, problems related to the testing and validation of complex software systems have grown exponentially.[31] For example, at the 1992 International Switching Symposium, Bellcore's Irwin Dorros commented that intelligent network functionality involves 'difficult and fundamental re-architectures of the software in many of today's switching systems...[and] re-architectures of the hardware itself' (Finnie 1992). Some PTOs argue that open network access and control by their competitors is a pipe-dream. They suggest that network access will evolve through the process of building cost-effective bridges between their own networks and those of their competitors. Thus concerns about network reliability and integrity cannot be dismissed as a repetition of historical events that resulted in monopolistic control of the telecommunication industry.

The final limitation of this study is that the development of the knowledge and skills base which informs the technical competencies of the suppliers of the intelligent network has not been examined in detail. Most telecommunication engineers argue that data processing and information transport require different knowledge sets. In this study, perspectives on this issue are reported by company representatives interviewed between 1989 and 1991. No independent assessment of their claims about the respective competencies of network designers in the computing and telecommunication field has been undertaken, and the extent to which these perceptions match the sophistication of software activities that are actually carried out within the PTO and equipment manufacturing firms has not been verified.

Nevertheless, the analysis in the following chapters shows how new modes of co-ordination between the telecommunication and computing firms are being devised to overcome the software bottleneck. The intelligent network offers an opportunity to European PTOs and their suppliers to introduce proprietary standards with each enhancement to existing plant (Bordeaux and Maher 1989; Kung 1989). Even where open interface standards are agreed, the internal complexity of the network can create entry barriers for equipment manufacturers and service providers.

## Network Access, Integration and Segmentation

It has been suggested by observers of the evolution of telecommunication networks that segmentation and diversity are creating a situation analogous to the emergence of two separate telecommunication systems. One is designed to serve the needs of the mass customer and the other is designed in response to the needs of multinational companies (Wieland 1988). Alternatively, network designs which integrate functions and services may provide a basis upon which to exploit economies of scale and scope which arise from the joint use of the public network by many services; for example, advanced imaging services and basic telephony. For instance, costly duplication of software functionality in public network switches and terminal equipment could be avoided as a result of an integration strategy (Slaa 1988; von Weizsaecker 1987).

Some argue that the growth of information service markets is suppressed by segmented network evolution or that flexibility in network use by small and medium-sized firms is reduced. There are concerns that suppliers and users will be locked into uneconomic long-term relationships with large PTOs or other network operating companies if a proliferation of proprietary standards for public and private networks emerges. The integrated public network approach is advocated by large as well as smaller firms. Large firms have requirements for the capabilities of advanced public networks as well as for private networks. Companies opting for investment in private networks often seek the advice and expertise of public network operators, and they make considerable use of public networks long after they have established private alternatives (Mansell and Sayers 1992). The strategic and policy issues relate, not simply to whether the public telecommunication network should be upgraded to perform 'intelligent' information-processing functions, but also to the implications of alternative designs for different types of users.

The intelligent network could provide a platform for a wide variety of ubiquitous services delivered throughout the public network by today's PTOs as well as a host of competing service suppliers. This is the basis for the Idealist model, and it is generally associated with a process of network segmentation. But there is also the possibility that the advent of an intelligent network will engender higher service costs for some users. This is the concern of the Strategic model, which sees an uneven, oligopolistic market for public and private telecommunication services unfolding. The costs of complex software essential to the intelligent network must be recovered in some way. The design choices discussed in the following chapters show how public and private institutions are shaping and reacting to these technical innovations and thereby establishing how the terms and conditions of network access will be altered.

A variety of incompatible segmented networks could be ushered in by the intelligent network. The RBOCs in the United States are implementing different versions of the intelligent network architecture. They are also resisting regulatory requirements to unbundle newly acquired network intelligence which would create more open access conditions. In Europe, the PTOs have yet to reach agreement on open interface standards for advanced electronic communication networks. Developments in the United States are examined in the following chapter, as this is where the generic architecture for the intelligent network first emerged.

## Notes

1 This refers to data processing and computer control techniques in the operation of a switch. The technique is generally referred to as Stored Program Control (SPC), and enables the use of computer programs to achieve centralized control over the operational, administrative and maintenance functions. It interprets signals received from subscribers and other switches, arranges the set-up and clearing procedures and paths through the switching unit and determines what messages are sent to the various subscribers or other switches connected to a particular switch. By the mid 1960s only a few SPC switches had been designed; see Meurling and Jeans 1985.

2 Freeman defines an incremental innovation as an improvement in a product or process using an established set of techniques defined within an overall technological paradigm (Freeman 1982).

3 Pulse Code Modulation (PCM) is a method of converting analogue signals (e.g. voice signals) to digital signals. The analogue signals are sampled and encoded to represent the signal levels of the wave form of the analogue signal in digital form.

4 In 1989, Bellcore announced an Advanced Intelligent Network (AIN) for 1995 which is intended to replace both IN/1 and IN/2. AIN was to be introduced in three steps: AIN Release 0 (1989), Release 1 (1992) and Release 3 (1995). Each of these was to be developed in co-operation with multiple equipment vendors (Magedanz and Popescu-Zeletin 1991).

5 R. Hawkins, Science Policy Research Unit, personal communication, January 1993.

6 The International Telecommunication Union CCITT produces recommendations within the framework of the International Telecommunication Regulations (1988). These recommendations do not have the force of law. A. Rutkowski (then Special Adviser to the Director General of the International Telecommunication Union) and G. Codding (Professor, University of Colorado) were the first to describe publicly CCITT recommendations as standards (R. Hawkins, Science Policy Research Unit, personal communication, January 1993). For a detailed discussion of the role of the International Telecommunication Union and other regional organizations in standards setting, see Hawkins 1992b.

7 It should be noted that in Europe the generic intelligent network terminology differs slightly from that adopted by Bellcore and AT&T. The European Telecommunication Standards Institute (NA6) refers to (WSF) Workstation Function; (SMF) Service Management Function; (SCEF) Service Creation Environment Function; (SCF) Service Control Function; (SDF) Service Data Function; (SSF) Service Switching Function; (SRF) Specialized Resource Function; (CCF) Call Control Function; (CCAF) Call Control Agent Function. Different elements can be allocated to the PTOs or to competing service providers, resulting in a definition of reserved or non-reserved (competitive) services; see Magedanz and Popescu-Zeletin 1991.

8 Common channel signalling is a method of providing control in a telecommunication network. Common channel signalling reduces the cost required for separate signalling units in each exchange and permits greater flexibility for the future development of services. Signalling System No. 7 refers to a CCITT recommendation for common channel signalling designed for use in networks conforming to Integrated Services Digital Network (ISDN), but not limited to this configuration. This system provides internal control and network intelligence but does not transfer data. This recommendation was developed during the 1970s. It defines the format and content of packets on the D-Channel in the ISDN configuration.

9 The TCAP (Transaction Capabilities Application Part) provides the network transaction capability in CCSS7. The TCAP uses the system of Signal Transfer Points and is the basis of signalling for various intelligent network components.

10 The 'D Channel' carries control signals and low-speed data and transmits at 16 or 64 kbit/s.

11 For a detailed analysis of the history and problems associated with the design and implementation of ISDN, see Darmaros 1992.

12 A Basic Rate Interface consists of two 64 kbit/s channels carrying information as 'B Channels' and one 16 kbit/s 'D Channel' carrying the signalling information or 2B+D. The Primary Rate Interface operates at 1.544 Mbit/s in the United States and consists of twenty-three 'B Channels' at 64 kbit/s each and one 64 kbit/s 'D Channel'. In the United States, the Primary Rate Interface for ISDN operates at the same speed as a T1 circuit commonly used in private networks. In Europe, the Primary Rate Interface operates at 2.048 Mbit/s and uses thirty 'B Channels' and one 'D Channel'.

13 CCITT provisional specification of intelligent network attributes includes the following capability sets: extensive use of information-processing techniques; efficient use of network resources; modularization of network functions; integrated service creation and implementation by means of re-usable standard network functions; flexible allocation of network functions to physical entities; portability of network functions among physical entities;

standardized communication between network functions via service independent interfaces; service provider access to the process of composition of services; service subscribers' control of subscriber specific service attributes; and standardized management of service logic; see Consultative Committee for International Telegraph and Telephone, International Telecommunication Union 1990.

Other relevant documentation includes CCITT Intelligent Network Recommendations, (Q 1211) Introduction to Intelligent Networks CS 1; (Q 1213) Global Functional Plane for Intelligent Networks CS 1; (Q 1214) Distributed Functional Plane for Intelligent Network CS 1; (Q 1215) Physical Plane for Intelligent Networks CS 1; (Q 1218) Intelligent Network Interface Recommendation CS 1; (Q 1219) Intelligent Network User Guide CS 1; (Q 1290) Vocabulary of Terms Used in Definition of Intelligent Networks.

14 These partnerships included Picture Tel, Video Telecom, Supercomputer Systems, Samsung Software, America Conductus, Northern Telecom and AT&T in the United States; and Bell-Northern Research (Canada); Telecommunications Research Laboratory (Denmark); Toshiba, Furukawa Electric (Japan); Industrial Technology Research Institute (Taiwan); and Siemens Aktiengesellschaft (Germany) (*Communications Week International* 1990).

15 At one extreme, the routing, billing and management of the network could be performed by suppliers other than the public network operator. At the other, the public network operators would provide all features and services. In reality, by the mid-1980s, a certain level of message processing was already embedded in the public network as part of the switching capability. The issue was how much greater the 'intelligence' of the public network should be and whether the RBOCs should be able to exploit this capability to add value to the basic conveyance of messages.

16 An institution is defined by W. H. Hamilton as 'a way of thought or action of some prevalence and permanence, which is embedded in the habits of a group or the custom of a people', cited in Melody 1987a: 1315. A more formal definition of an institution is as the embodiment of recognizable procedures that mediate the interactions among members of groups or collectivities within society, or in society as a whole, see Berger and Luckman 1966.

17 The institutionalist economics tradition grew out of the work of Veblen and others, such as John R. Commons, W. Mitchell and J. M. Clark. It placed the study of institutions at the centre of research on economic problems and was critical of theories and models that ignored the dynamics of institutional change, power relations and historical patterns. This tradition was also concerned with the ways in which public policy could be used to eliminate monopoly power and special privilege through reforms and regulation; see Gray 1981; Melody 1987a. More recent developments of formal approaches to the analysis of institutional change are reviewed in Hodgson 1988.

18 Carlsson and Stankiewicz define a technological system as follows: 'a network of agents interacting in a specific economic/industrial area under a particular institutional infrastructure or a set of infrastructures and involved in the generation, diffusion and utilization of technology' (1991: 111).

19 The concept of bias in the macro-level structure of technical change in the information and communication environment is developed in the work of Innis; see, e.g. Innis 1951; Melody et al. 1981. Concepts of surveillance and control leading to bias at the micro-level are addressed in Giddens' work; e.g. 'Information storage is central to the role of "authoritative resources" [i.e. resources that allow control over the socially constituted world] in the structuring of social systems spanning larger ranges of space and time than tribal cultures' (Giddens 1985).

20 There are other institutional aspects of the design process which are important but are not the focus of this study. For example, the industrial research and development system in each of the countries under investigation is not examined. Similarly, the implications of change in the wider institutional environment – macro-economic and competition policies – are considered only in so far as they bear directly on the design priorities for the intelligent network.

21 A technological paradigm is an exemplar or an artefact that is to be developed or improved. It is also a set of heuristics which defines characteristics of the bundling of techniques. Problem-solving models are 'rules on how to search and on what targets to focus,

and beliefs as to "what the market wants" become the shared view of the engineering community'. A technological trajectory is 'the activity of technological process along the economic and technological trade-offs defined by a paradigm' (Dosi 1988: 1127).

22 For example, Dosi argues that 'the endogenous nature of market structures associated with the dynamics of innovation, the asymmetries among firms in technological capabilities, various phenomena of non-convexivity, history dependence, dynamic increasing returns, and the evolutionary nature of innovation/diffusion processes are some of the main elements of the process of technological change' (Dosi 1988: 1164).

23 Studies in the political economy and institutionalist traditions have focused mainly on the structure and role of policy and regulatory institutions and on the formation of strategic alliances among firms. The standards-setting literature has only begun to tackle the telecommunication field and has yet to link standards-setting processes with the literature on technical design, innovation and paradigmatic change. Exceptions are Hawkins 1992b and Mansell and Hawkins 1992. Research on the formation of socio-technical constituencies in the information technology-sector generally and in computing and telecommunication by Molina has given consideration to technical design issues in a way that complements the present analysis, see for example Molina 1989.

24 A design configuration defines

> precisely the purpose, mode of operation, construction materials and method of manufacture of the relevant artefacts. It defines the evaluation framework for the technology and it is within the configuration that specific design puzzles emerge . . . the notion of a design configuration plays a fundamental role, allowing the distinction to be made between competition *within* a design configuration and competition *between* design configurations. (Metcalfe and Boden 1992: 60–1; emphasis added)

25 The concept of a productive transformation unit is closely related to the notion of an architectural innovation. For example, Henderson and Clark suggest that 'the essence of architectural innovation is the reconfiguration of an established system to link together existing components in a new way' (1990: 12). The evolution of the intelligent network involves more than the technical reconfiguration of existing components, and the choices with regard to architectural innovations need to be considered in the light of changes in the institutional environment.

26 This perspective is similar to the expectation that innovations in the architecture of a technical system will create opportunities for firms which introduce the innovation to gain advantage over well-entrenched firms (Henderson and Clark 1990: 28).

27 For example, in 1989, the percentage of total subscriber lines attached to digital switches was 70.7, 38.0, 42.5 and 2.6 respectively, for France, the United Kingdom, the United States and West Germany. The percentage of total exchanges with CCSS7 capability in the same year was 47.6, 6.2 and 6.3 percent, respectively, for France, the United Kingdom and the United States (*Communications Week International* 1991b).

28 For example, for some services only the capacity to store and translate numbers – e.g. freephone – is required. For other interactive services – e.g. voice messaging, call completion, private transaction network and pay-per-view – more sophisticated network access is needed.

29 A well-known example is the early goal for the United States' Bell System, which was articulated by Theodore Vail. 'The telephone system to give perfect service must be one in which all parts recognize a common interest and a common subordination to the interests of all, in fact it must be "One System", "universal," "intradependent," "intracommunicative," and operated in a common interest. Such is the Bell System'. See AT&T's 1916 Annual Report, pp. 40–1, cited in Trebing 1969a: 305.

30 Interviews with Swedish Telecom and DBP Telekom, February–March 1990.

31 In October 1991 these concerns had come to a head with the failure of CCSS7 implementations in the United States, Japan and Sweden. A single software problem in one Signal Transfer Point had shown that it could wreak havoc in the public network. A meeting called by the United States Federal Communications Commission, and convened by the CCITT, led to an agreement to exchange information with respect to faults on a voluntary basis

among PTOs. In September 1991 AT&T's network in the United States was shut down, the third major switch failure in less than two years. Other failures had occurred in the United States despite the fact that only about 20 percent of the RBOC's networks were employing CCSS7. In the case of Bell Atlantic, the problem was traced to equipment made by DSC Communication Corporation of Texas, but in each failure the network condition which created the problem was different. In the United States, some of the PTOs, such as MCI and Sprint, have been allowing their corporate customers to access their CCSS7 out-of-band signalling networks directly, and in the eyes of some, this is creating even greater risk of network disruption and threats to security (Lynch 1991b; Lynch 1991c; Lynch and Herman 1991; Sweeney 1991).

# 3

# Early Network Transformations – the US Experience

The mirage was called 'convergence'. It grew out of the perfectly correct observation that, thanks to the microchip, the basic technologies used in data processing and telecommunications were fast becoming identical. Hence the two businesses appeared set to merge into a single information industry, which would amalgamate both disciplines.

(de Jonquieres 1989)

**'Reach Out and Touch Someone'**

During the 1980s the 'mirage of convergent technology' contributed to a belief that the innovation and diffusion of advanced digital and software technologies would coincide with the demise of the power of monopolistic suppliers to control the design and use of the telecommunication infrastructure (de Jonquieres 1989). Telecommunication supply would come to approximate the conditions of the Idealist model of network evolution.

The road towards technical convergence in the United States was initiated during the era of the analogue telecommunication infrastructure, when digital technologies were not even available for commercial use in the public network. In fact, the first resonance of competitive entry began in the late 1940s and early 1950s with the stirrings of change in the terminal equipment market. Recording devices were coming onto the market and the Hush-a-Phone was developed to allow a caller to shield sound from nearby listeners.[1] AT&T argued that this device would damage the technical integrity of its network.

Digital voice circuits began to be used in the US telephone network as early as 1961 to multiplex voice circuits (Solomon 1991). As a result, competitive entry was associated with digital technologies. Digital voice circuits had been invented and implemented during World War II to encrypt voice signals. The digital circuits used in the early 1960s were T-1 carrier links used by Western Electric. Competition was soon to be regarded as an inevitable response to technical change. In network transmission, AT&T was challenged in the late 1950s by private telecommunication users who won access to the radio frequency spectrum for non-common carrier microwave services.[2] In 1969 AT&T was unable to bring a halt to MCI's construction of a single microwave link between Chicago and St Louis.[3]

Some observers of the changes in the market argued that the conditions embraced by the Strategic model of telecommunication development would bring a continuing need to protect consumers from the market power of dominant companies in the telecommunication market (Selwyn and Montgomery 1987; Shepherd 1969; Trebing and Melody 1969). Although the names of the dominant companies might change, concentrations of market power and the ability to exploit it would not vanish in the future. The transition from the monopolistic era to a more competitive environment would not be without its problems.[4]

Despite almost two decades during which it had been challenged by competitors, AT&T was still the dominant player in the US telecommunication market at the beginning of the 1980s (Melody 1986). An AT&T advertising phrase of the period, 'Reach out – reach out and touch someone', was in one sense merely a slogan to stimulate telephone-call revenue growth. However, by reflecting AT&T's ability to exploit access to telephone subscribers through the Bell Operating Companies, which AT&T owned and controlled, the slogan also became a beacon for those who claimed that AT&T would employ anti-competitive tactics to retain its dominant share of the market.

This chapter looks at the ways in which the new institutional structures led to an innovative network design in the United States. The detailed history of the changes in the market and in the regulatory regime has been recounted elsewhere (see, for example, Gabel 1969; Hills 1986; Temin and Galambos 1987; Trebing 1969a), and the purpose of this chapter is to examine the issues which arose in the mid-1980s with regard to access to the public network infrastructure. The focus is upon how these issues influenced, and became embodied within, the technical design of the public network.

## Policies and Strategies, Structural Separation and Open Networks

Two of the companies seeking to enter the American voice telephone market during the 1970s and 1980s, MCI and Sprint, found it necessary to challenge AT&T in the courtroom. The charges included the failure of AT&T to interconnect the new network operators, the conditions and prices for interconnection – for example, line versus trunk side interconnection – and allegations that AT&T's various service-pricing schemes were predatory. It was argued that AT&T's various abuses of market power should be regarded as unjustified attempts to stamp out competition before there had been a chance to demonstrate that market competition could be effective in stimulating greater efficiency of supply by AT&T, the incumbent in the market. The slogan used by MCI at one point in its challenge to AT&T was 'Reach out – reach out and crush someone', indicating the new entrant's concerns about the barriers to entry which AT&T raised as the company sought to build market share.

As a result of a series of challenges to AT&T that were considered by the Federal Communications Commission (FCC) and the courts, AT&T finally agreed to a consent decree, labelled a *modification of final judgment*.[5]

Negotiated between the Department of Justice and AT&T, this judgment called for the divestiture of various components of AT&T's network. The result in 1984 was the establishment of seven new regional Bell Holding Companies which came to be known as the Regional Bell Operating Companies or RBOCs,[6] and a *new* AT&T which no longer controlled the local exchange 'bottleneck' facilities in the public network. With this massive restructuring and reallocation of assets, the view that the market power of AT&T had been broken became even more pervasive, allowing the Idealist model to gain new adherents.

At this point in history, the main issue for this model's adherents became that of considering how AT&T would fare in the restructured marketplace (Pisano et al. 1988; Teece 1989). Few of the analysts who were concerned with the innovative capacity of the US telecommunication industry or with the competitive challenges that were on the horizon in the global telecommunication market observed that the domestic marketplace had virtually none of the hallmarks of a highly competitive market. In fact, it had retained a substantial number of distortions.

The RBOCs, for example, controlled the vast majority of the local and intrastate public telecommunication market within their respective territories, although the latter amounted only to about one quarter of the total long-distance market in the United States. The judgment had prevented the RBOCs from entering advanced information service markets, but they had control of the local exchange telecommunication infrastructure 'bottleneck'. Additionally, AT&T retained its dominant share of the long-distance telecommunication market even though it operated under restrictions.[7] The division of the 'old' AT&T's assets had been devised as a pragmatic solution to the need to change the economic incentives that had operated in the market before the 'break-up'. For the Idealists, the best way to ensure that effective competition would take hold after the divestiture agreement was implemented would be to withdraw regulatory restrictions in order to allow a free reign to market forces.

In the mid-1980s, market analysts began to consider the implications of new forms of competitive telecommunication markets. Few contemplated the implications of what was, in reality, rivalry in very imperfectly competitive markets, or the problems this would create for public policy. The high-profile spokesman who bore the burden of countering the Idealist vision of the market was the United States District Court judge who had presided over the break-up of AT&T.[8] Judge Harold Greene argued that divestiture had merely created new ground rules for the evolution of telecommunication in the United States – it had not diminished the opportunities for AT&T or the RBOCs to act against the public interest in telecommunication. From 1984 onward, his opinions on the telecommunication market were roundly criticized by AT&T and the RBOCs, who argued that the very act of restructuring had made all the telecommunication network operators answerable to their customers in the marketplace. The RBOCs petitioned frequently for 'line of business waivers' which would permit them to enter into equipment manufacturing and information-service provision from which they had been banned by the 1984 divestiture agreement. In the first Triennial Review of the agreement, the report prepared for the Department of Justice argued that neither AT&T nor the RBOCs could control the development of the telecommunication market in the United States (Huber 1987). For example, in summarizing the changes in the US telecommunication industry in the 1987 review of the AT&T divestiture agreement, Huber argued that,

> In the last fifteen years a revolution in technology has transformed the way in which network engineers think, plan, and build. The network has changed from a pyramid to something more like a geodesic dome. . . . That neither IBM nor AT&T has come close to capturing the market for electronic intelligence is attributable, once again, not to regulation or antitrust law but to the underlying technology itself. And here, both technology's successes and its failures have worked together to thwart any would-be monopolist. (Huber 1987: 1.2, 1.26)

Many feared that the United States District Court had become a regulator, and, at the same time, the FCC sought to re-establish its authority as the regulator.

The FCC began to champion the view that AT&T was no longer a dominant carrier in the US domestic market.[9] During the challenges to AT&T's monopoly in the 1970s and early 1980s, the FCC had vacillated between supporting and defending the incumbent's monopolistic control of the telecommunication network. AT&T claimed that its actions in the market and its treatment of competitors stemmed from the need to protect the integrity and ubiquity of the public network. When competitive entry was permitted by the FCC or the courts, it was not long before arguments ensued over the terms and conditions of access to the public switched telecommunication network.

Pricing strategies, proprietary standards for interconnection of competing networks and customer demand for network integrity were the tactics that AT&T used to rebuff challenges to its market.[10] These became focal points for policy-makers and new market entrants in their battle to push back the monopolist's control of technologies and markets. It was in this environment that the intelligent network came to the fore as an architectural concept which might enable competitors in the telecommunication market to take advantage of the benefits of technical change and innovation in service supply.

The original intelligent network concept was, in some respects, simply a marketing strategy devised by the RBOCs much as the call to 'reach out and touch someone' had been for AT&T. For example:

> Pacific Bell describes the intelligent network in terms of an 'Information Age vision' in which 'the Goal is simple – to bring Information Age services and benefits within easy reach of everyone already served by the local telephone companies by offering services, and through lowering barriers to participation and use for thousands of providers and millions of potential customers. (Pacific Bell 1987)

The idea of a new generation of public network designs which would be responsive to the customer's desire to gain greater control of the costs and functionality of a vast new array of services strengthened the impression that competition was in full swing. Not only did the concept fit like a well-made glove with the Idealist model of telecommunication, it could also be used to support arguments in favour of the removal of the heavy hand of regulation. It was regulation that was held responsible for the failure of AT&T to innovate in response to technical change. The idea that AT&T's failure to bring advanced services to the market might have been linked to its vast size and monopoly position in the market was rarely countenanced by advocates of the Idealist model of the marketplace.

However, the intelligent network concept was not simply a marketing strategy. From the telematics platform it would create, the RBOCs planned to launch themselves as providers of advanced information services. In other words, they sought to position themselves for the day when they would be free from restrictions imposed at the time of their divestiture from AT&T in 1984.[11] These restrictions were regarded by the RBOCs as anathema. They worried that they would become *bit transporters* and, despite the fact that they were already beginning to offer increasingly advanced services such as freephone, virtual private networks, automatic call distribution, alternative billing services, wide area Centrex, emergency number routing and incoming and outgoing call management services, they argued that they would be locked out of the market for the development of the advanced information services that the 'intelligent' public network could potentially support.

Most of these advanced services mimicked those already offered by the long-distance (interexchange) carriers such as AT&T, MCI and Sprint. The RBOCs argued that the intelligent network would enable them to be more responsive to demand, and to reduce the time and cost of configuring equipment to support new services. The key to this new world was to be the introduction of complex software-based systems (Solomon 1991).

In the United States at the beginning of 1987, the Idealist vision of a 'universal information networking service' was becoming strongly entrenched in technical and public-policy circles. This vision was, for example, to allow for an environment in which 'today's geographical addressing and switch-based routing will be superseded by geographic independence. Addressing and routing will be [orientated] to individuals, not to a geographic location' (Dorros 1991: 3). Gigabit-per-second optical fibre systems and terabit-per-second packet switching would combine with network intelligence. However, there was a risk. Both AT&T and the RBOCs still had billions of dollars invested in their more conventional pre-intelligent networks. Were they in a position to introduce strategies to ensure that their shareholders would not bear the costs of early obsolescence? The answer depended in part, on the degree to which their networks would be openly accessible to new types of users, including their competitors. It also depended upon who would bear responsibility for a more rapid depreciation of existing plant than had been planned prior to the advent of the new 'competitive' era.[12]

The question of the degree to which the public intelligent network infrastructure could be open to access by intermediate and end-users has occupied centre-stage in telecommunication policy and regulation in the United States. Even before divestiture, the structural separation of telecommunication operations in the monopoly (generally, voice telephone) domain and in the competitive domain had preoccupied regulators.[13] But structural alternatives able to contain the market power of the RBOCs and AT&T were not the only regulatory tools at hand. Technical designs which could create 'Chinese walls' between monopoly and competitive activities were also under consideration.

The RBOCs, AT&T and their competitors might be able to develop a vision of a new intelligent infrastructure. But turning the vision into reality would require that the equipment manufacturers be involved in its design and implementation. Certainly this was the view of the manufacturers. AT&T, Northern Telecom and a few other companies saw the need to ensure that their designs complemented the requirement specifications of the network operators in the US market. This strategic view of design was important because, by 1987, the presence of manufacturers from outside North America could be felt in the domestic terminal equipment market. The American- and Canadian-owned equipment manufacturers felt that competition for products used in the core of the telecommunication infrastructure could not be far behind. Spurred on by the view that convergence of computing and telecommunication technologies was synonymous with convergent manufacturing empires, the manufacturers sought to establish their credibility in the new multi-vendor environment of the 1990s.

It is clear from this brief analysis that it was the dynamics of the Strategic model that shaped the responses of companies in network, services and equipment markets. The recent history of the strategies and tactics used by AT&T and the RBOCs illustrates how the intelligent network failed to become the harbinger of competition as promised by the Idealist model.

## Designing Intelligent Network Architectures

### *The Original Approach of the Regional Bell Operating Companies*

As a result of the 1984 divestiture, the RBOCs claimed that AT&T had acquired all the Bell System's 'intelligent network' equipment.[14] Prior to divestiture, AT&T's practice had been to introduce 'intelligent' functionality to support advanced services in the long-distance portion of the public network first. Consequently, the RBOCs argued that they would be forced to develop their own intelligent services if they were to gain a foothold in new markets. By 1987, the FCC had devised a regulatory regime which would allow the RBOCs to offer some advanced information services providing they implemented Open Network Architectures (ONA). In this way the FCC sought to ensure that RBOCs could not abuse their monopoly control of the local exchange portion of the network.

The RBOCs recognized that the most significant revenues would accrue to the operators responsible for maintaining the databases in the new intelligent network environment (Korzeniowski 1988b). The issue was who would gain control and economic supremacy in this new market. Credit-card calling services had been developed by the RBOCs and AT&T, but these depended on AT&T's database. The initial intelligent network design would allow other suppliers to control and manage databases for their own services. Revenues would be generated by surcharges on card usage and by query charges for card validation using a database. This pattern was to be repeated for a host of other services using intelligent databases.

The RBOCs also recognized that the intelligent network offered a way of stimulating network traffic. They planned to use virtual private networks to challenge customer equipment-based private networks by relieving the user of responsibility for network management and maintenance. The features associated with virtual private networks included abbreviated dialling, authorization codes, call forwarding and screening, call identification, private numbering and site location codes, switched data transmission and a variety of other network applications. Many of the RBOCs played on user fears that technological change would render obsolete customer-owned facilities in private networks which already supported some of these features.

In response to the RBOCs' initiatives to develop a more advanced network infrastructure, AT&T petitioned the court to bar the RBOCs from providing virtual private networks, claiming a violation of the AT&T divestiture agreement. AT&T argued that it had the same type of architecture as that being proposed by the RBOCs and that the companies had been constrained from competing in one another's markets by the agreement. Furthermore, in many states, the local exchange monopolies had been retained as a result of legislation or regulatory policies. AT&T was concerned that the operating companies would have incentives to recover the costs of investment in advanced network infrastructure using revenues generated by their local service customers, the majority of whom would be unable to opt for alternative service supply. In the long-distance market, AT&T could no longer access this pool of revenues as it had prior to the divestiture of the Bell Operating Companies.

To implement the intelligent network the RBOCs needed Common Channel Signalling System No. 7 (CCSS7), as described in Chapter 2. Out-of-band signalling capability would allow information related to call processing and routing to be communicated throughout the public switched network. But the CCSS7 design had to be licensed from AT&T, which had developed the version that operated within the US public network, and the licensing fees were a cause for concern among the RBOCs. The total cost of upgrading was estimated to be in excess of US$40 million.

The RBOCs also needed to tap every possible source of revenue if they were to recover the costs of a massive network modernization programme. To achieve this, the RBOCs introduced local premium (advanced) services. The prices for these services might be increased as long as they retained their local monopolies.[15] One set of services, CLASS (Custom Local Area Signalling Services) for residential users, included call screening, call tracing and call blocking.

Bell Atlantic was the first RBOC to convert to CCSS7 in 1988 in order to support intelligent network services.[16] Soon after, Bell Atlantic announced plans for the intelligent network. BellSouth and US West were moving ahead with intelligent network trials. Although investment in pre-intelligent and intelligent network services had been underway for some time, by 1991 less than 1 percent of residential customers had access to the CLASS services which were supported by the intelligent network capability (Robrock 1991). In 1993, many areas in the United States still could not access these services.[17]

By early 1990 the RBOCs began to warn that, unless they were freed to offer advanced information services, the intelligent network would not be built. They argued that investment – namely, digitalization and CCSS7 – in the United States was lagging behind that of other countries. The 1984 divestiture agreement had barred the RBOCs from offering virtually all types of information services. In 1988 Judge Harold Greene had permitted them to provide *transport only* services. The RBOCs appealed this ruling, and in April 1990 the United States Court of Appeals for the District of Columbia returned the original ruling to Greene for further consideration and the restrictions were ultimately removed. The FCC welcomed this, and said it intended to use ONA and other non-structural safeguards to ensure that the RBOCs opened access to their networks to competitors.

The ONA provisions had been developed originally during the FCC's 1986 Computer Inquiry III.[18] Open Network Architecture (ONA) is a regulatory concept promulgated by the FCC which requires the Bell Operating Companies to offer access to basic service elements (BSEs) for the provision of enhanced services. Implementation of ONA enables the companies to offer certain enhanced services without structural separation of these activities from the provision of basic services. Basic services are limited to services providing a transparent communication path. The provider cannot control or manipulate the contents of transmissions. Basic services include: wide area telephone service, private line, video transmission, and services closely related to basic services such as abbreviated dialling, directory assistance, billing and voice encryption services. Enhanced services are all other services – for example, where the provider uses a computer to act on the format, content, code, protocol or similar aspects of the transmitted information, furnishes the subscriber with additional, different or restructured information, or involves the subscriber's interaction with stored information. The controversy as to the stability over time of these distinctions and the category into which specific services fall has bedevilled both the FCC and state regulatory commissions.

In 1988 the FCC approved blueprints for the ONA model proposed by the RBOCs. The models offered enhanced service providers access to the RBOCs' underlying network services called 'basic service elements', or BSEs, via access links called 'basic serving arrangements' (BSAs). These network resources were to be offered under federally regulated tariffs. The FCC rejected arguments by competing information service providers, who claimed that they should be allowed to choose which parts of the public network infrastructure they wanted to use. In consequence, the RBOCs retained the ability to design the access conditions under which they would compete with new entrants. It was not until 1991 that the FCC ordered the RBOCs to submit tariffs for access, and these took effect in 1992.

Clearly, the lesson of the mid-1980s is that the substantial market restructuring, resulting from the AT&T divestiture agreement, had not automatically opened the door to the market conditions promised by the Idealist model. For example, the RBOCs saw the design of the intelligent network as a way of exercising control over the services they would provide once they were free of the divestiture restrictions.[19] They did not wish to remain dependent on generic software packages offered by switch manufacturers. The RBOCs needed service-creating capability, and this was the priority. Only in the longer term would the databases used to create and manage services be available to larger customers or residential users to enable them to design, control and manage their own services.

The RBOCs' intelligent network freephone service was introduced as a result of their desire to compete with AT&T.[20] In 1986, the FCC had ordered equal access for freephone services. The RBOCs had been using a three-digit code to identify these calls. By 1988, a ten-digit code to identify the service provider of a freephone number had been devised. The result was that freephone service provision was freed from its dependence on exchanges controlled by a single operator (in this case AT&T), and a step was taken towards the independence of these services from the physical configuration of the network.

The RBOCs also planned to offer more sophisticated time-of-day and least-cost-routing services for their larger customers. These services were being offered by AT&T and other long-distance carriers, and the RBOCs argued that they would fall behind if they did not invest rapidly in equivalent intelligent network capabilities. The battle between AT&T, the other long-distance carriers and the RBOCs continued over the introduction of services such as alternative billing, credit-card validation and Centrex. All these services were part of the tool kit that AT&T, the RBOCs and other facilities-based carriers would use to draw private corporate users to the services offered using the public switched network-based services.

The RBOCs' network upgrading plans to introduce what they described as the 'intelligent network' were in full swing by 1988. Pacific Bell introduced 'Netsys' in October 1988. This was a telematics service platform which was closely associated with the intelligent network. AT&T and Northern Telecom responded with equipment offers, but both these manufacturers were concerned about the need to design new open interfaces which would conform with ONA and other regulatory specifications. Both manufacturers were already developing proprietary systems to support the intelligent services that AT&T and the RBOCs argued would be needed to meet customer demand in the coming decades (Pacific Bell 1988).

*The AT&T Response*

Two divergent approaches to the intelligent network had emerged by 1991. First, there was the RBOC version which provided voice services to residential customers – such as freephone customer request services – with the aim of stimulating traffic volumes. Second, AT&T promoted an alternative approach based on the introduction of a universal information system which would generate long-distance traffic, especially among the large, corporate users with requirments for voice and non-voice services. Both strategies have resulted in the upgrading of the signalling capacity of the public network.

Since the 1960s, many of AT&T's largest customers have shifted their traffic to corporate networks based upon private lines provided by the company at very low prices. For example, the TELPAK tariff was introduced in 1961 as a quantity discount for private lines. Discounts ranged up to 85 percent as compared with the traditional private-line tariffs, which were, in turn, discounted from public switched network service tariffs. These discounted lines could be combined with reduced prices and common control switching arrangements (CCSA) to establish private networks (Federal Communications Commission 1967).

In 1988 it was claimed that US$3.7 billion was lost to bypass traffic. Bypass is defined by the FCC as communication traffic that originates or terminates without using the facilities of the local exchange (OECD 1988: 28). Bypass is closely associated with the growth of private networks. Private network facilities in the United States can be owned by companies, or based on leased lines owned by the facilities-based carriers. Private networks include those which are operated by larger companies using the leased circuit capacity purchased from a public network operator. Private networks can use entirely separate physical facilities – for example, private satellite capacity or a company's own dedicated terrestrial microwave capacity. Internal private corporate communication links that improve co-ordination and management were being extended to embrace large firms' suppliers and customers. The aim was to achieve substantial time and cost savings. To satisfy their communication needs, large firms often combined their private networks with the public network (Bar and Borrus 1989).

Private networks were not a new development in the 1980s and the 'competitive era'. Historically, the special requirements of some users had been met through self-supply or 'private' networks (Mansell 1986). In fact, even before competitive network operators had become established in the United States, it was AT&T which was the greatest proponent of bypass solutions. Prices for leased lines were kept low often below their direct costs, in order to provide businesses with a lower-cost alternative to the public switched telecommunication network (Melody 1977; Melody 1989; Trebing and Melody 1969; Trebing 1969a). It was when AT&T, and later the RBOCs, saw the first inroads into their market that the threat of bypass was used to argue that regulated competition would lead to the segmentation of the public network infrastructure and to the erosion of the systemic integrity of the US public telecommunication network.

The important consideration in the development of private networks was that configuration, access and applications could be controlled to some degree by the private network operators who did not need to rely entirely upon the public network operator to meet their service requirements (Bar and Borrus 1989). T1 (1.544 Mbit/s) circuits had been used by AT&T in the early 1960s to connect switches in the public network. By 1983, T1 circuits were being tariffed for private network users. By 1989, in an attempt to attract large telecommunication users who had established private corporate networks using bypass facilities back to the public switched network, long-distance carriers such as AT&T, MCI and Sprint were developing enhancements to their switched and unswitched services in order to regain control of traffic and revenues.

The overall digitalization of the telecommunication network in the United States between 1980 and 1990 had increased substantially.[21] The greatest change was in the inter-city long-distance market which, until the early 1980s, had been supplied almost exclusively by the Bell System. In 1989, MCI announced the implementation of Digital Access Cross-connect Systems (DACS) to offer real-time network management. AT&T announced Accunet at about the same time, a product which would allow customers to monitor their networks and reconfigure them in real-time using a Customer Controlled Reconfiguration system and a Bandwidth Management system.

But this was not the only facet of AT&T's response to competitors. The first pre-intelligent network concepts had been under development since the end of the 1970s. In the early model, the interface between the Service Switching Point and the Service Control Point was proprietary and service-specific. So, too, was the Service Management System – NETSTAR (Network Service Transaction and Recording System).[22]

The company announced in 1989 that it would spend some US$2.5–3 billion over the next few years to upgrade its network. The key elements in the network modernization programme would be intelligent network gateways and databases. Features were to include self-healing networks, automatic reconfiguration in the face of service disruption, dynamic bandwidth allocation, improved network management, and integrated voice and data applications. AT&T was expecting to complete the implementation of CCSS7 by the end of 1989, enabling it to provide transparent user-to-user information services and clear channel capability for data services.

In 1989, MCI and Sprint had already implemented CCSS7, and these companies were advertising intelligent network services especially for corporate users. AT&T countered by pointing to the fact that it had been offering intelligent network services since 1976 when it had installed a Common Channel Signalling System (CCSS) based upon packet switching to control all AT&T calls. This signalling system was based on the CCITT CCSS6 protocol, the predecessor to CCSS7, and it did not permit out-of-band signalling, the main element of the intelligent network architecture configuration.[23]

AT&T introduced the Software Defined Network, a domestic, virtual private network, in 1986, which made use of the enhanced 'intelligence' in its network. By 1989, the company claimed to have 110 customers connected via 500 serving offices. The Software Defined Network was a non-dedicated private line service which was marketed as a service with higher reliability and lower costs than could be achieved using leased lines. Marketing focused on the greater degree of flexibility, easier upgrading and centralized operation, billing and maintenance that AT&T could provide. In 1989, AT&T also introduced Express Connect, a service which enabled users to dial a seven-digit number to connect to any location and to establish point-to-point private line services. In October 1989, AT&T announced the *2000 Product Family*, which pointed the way to the development of new software for the company's 5ESS (Electronic Switching System) public network switch to support intelligent networks (Technology Investment Partners and Palo Alto Management Group 1990). The 5ESS switch had been launched in 1984. Its architecture allowed processing power to migrate to peripheral modules in the switch. About 75 percent of the design work that went into this development was said to relate to software. Until the late 1980s, hardware vendors like AT&T had supplied virtually all the software. AT&T argued that the public network operators wanted to be in a position to perform software modifications themselves in order to speed up the introduction of new services. Nevertheless, AT&T intended to remain the dominant supplier of software for its equipment. In 1989, the company was spending more than US$1.7 billion a year on equipment-related software development (Oram 1989).

In 1990, AT&T announced that AT&T Paradyne, a subsidiary, would supply the AT&T Accumaster Integrator software to support network management in the AT&T Unified Network Management Architecture. This product would compete with IBM's NetView and DEC's network management products, putting the company into a stronger position to manage large corporate computer networks.

All these products came together to form AT&T's intelligent network strategy, which was to offer virtual network services that would support multi-point features using the public network. In this way, the company hoped to hold onto its existing large customer base and to extend its market into new areas. As AT&T suggested,

> a natural extension of VPN [Virtual Private Network] is Centrex, which provides PBX-like features. A full VPN/Centrex service can meet all or most telecommunications needs for many businesses, *without the need for any privately owned and*

> *maintained equipment*. In addition to replacing private networks, the IN-based VPN service can offer services to those who would not traditionally have used a private network. (AT&T 1990: 13; emphasis added)

AT&T's plan was that, as enhanced 5ESS capabilities spread throughout its long-distance network, customers would be able to extend Centrex services from city-wide to nation-wide coverage. For example, a company, or group of franchises in different locations, would be able to form one Centrex group and obtain the advantages of a single internal PABX-based private network. Virtual private networks would support the networking of large and small PABXs. Centrex services would become attractive to almost any organization which possessed multiple locations. This strategy would allow AT&T to address a market segment that was being encroached upon by private competitors (AT&T 1991).

AT&T's flexible intelligent network made it increasingly feasible to tailor prices for services which configure special bundles of functionality to support the requirements of its largest customers. For example, Tariff 12 was first introduced in 1985 for the United States Defense Department network. In April 1987 this tariff was extended to other commercial users, the first of whom was the General Electric (GE) Corporation. Apart from General Electric, users included Ford Motors, American Express, American Airlines, Federal Express and Allied Signal. The FCC rejected Tariff 12 in April 1989, arguing that it was anti-competitive. However, an FCC spokesperson suggested that if AT&T could show that it incurred different costs in serving different users, it could charge different rates (Samuel 1989). AT&T redesigned the tariff and was subsequently allowed to implement it.

Other substantially discounted tariffs soon followed as AT&T sought to strengthen its hold on its large customer base. For example, in March 1988, AT&T introduced volume discounts for all its domestic and international 800 freephone services. Tariff 15 was described as a competitive pricing plan. Under this tariff AT&T could provide customer-specific volume discounts by designing tariffs to undercut those on offer by MCI and Sprint. After investigation by the FCC, the tariff was permitted to go into effect in September 1988.

Another illustration of AT&T's ability to respond to inroads in its market was Tariff 16, originally a Government Outbound Calling Plan. In December 1989 AT&T offered this tariff to the University of Texas and to the Pennsylvania Department of General Services. Similarly, in 1989 AT&T filed an international custom-designed network tariff for its International Virtual Private Network service initially for GE International and the Prudential Insurance Company. This service was called Virtual Telecommunications Network Service (VTNS). GE had signed a five-year agreement with AT&T, British Telecom International, and France Télécom Inc. to build a private network, and this was the first time AT&T had designed an international private network for a corporate user.

To provide these intelligent network services, information on the service features needs to be entered into a database. In the AT&T model, this process is supported by the NETSTAR service creation, provisioning and administration system. This allows the creation, updating and deletion of individual numbering plans, related service tables and associated administrative data by both the network operator and the customer. However, network traffic management under the control of AT&T is essential to the design. Described as a management system that can be used to control traffic routed to the Service Control Points during periods of network overload, by its operation AT&T could use this system to control access to the network by its competitors. Similar systems were being introduced by the long-distance interexchange carriers such as MCI and Sprint.

### *Northern Telecoms View*

In the US market, AT&T's equipment manufacturing activities had been challenged successfully by Northern Telecom, which extended its American operations significantly following the AT&T divestiture agreement. Northern Telecom is at the centre of the Canadian telecommunication equipment industry.[24] Its rise to prominence was the result of its innovative R&D performance and the relatively low-cost switches which it marketed in the United States.[25] Northern Telecom has been Bell Canada's preferred supplier and this relationship gave the company access to information vital to equipment design.[26] As the company put it: 'Investment risks are minimized, advantage is taken of production economies and product evolution, and development costs are spread over a longer time horizon and larger production volumes. As well, the revenues generated provide Northern with funds for future R&D investments' (Northern Telecom 1987: 5).

Like AT&T, Northern Telecom had been actively designing a new generation of intelligent network products from the mid-1970s. Northern Telecom introduced the intelligent network concept Vision 2000 in 1986, ten years after committing itself to digital products with its *Digital World* announcement. In 1977, the company had launched the digital 'revolution' in the United States by selling the first digital public switch before AT&T's equivalent product reached the market. In 1982, Northern announced the *OPEN World* (Open Protocol Enhanced Network) concept, indicating that the company had recognized that information systems would need to be compatible since no single vendor would ever be able to meet all communication needs (Northern Telecom 1987).

By 1988, the company claimed to have achieved its goal of embodying open interfaces in all its equipment, and this was regarded as the basis for future development of intelligent networks (Northern Telecom 1988).

> Today's advances in transmission, switching, and signalling represent the metamorphosis of the telephone network, taking it far beyond conventional connectivity and carriage in its economic and social utility. The intelligent network will be an electronic gateway to a busy array of services that respond to growing

> public need for: better education and training; better medical care; and better data communications. (Northern Telecom 1988)

In the 1988–90 period, Northern championed the idea that multiple switch manufacturers could collaborate in the design and manufacture of equipment for public and private telecommunication networks. The company began to shift its emphasis from hardware to software development, and it became increasingly dependent on external suppliers in the United States. The dominant view in the company favoured convergence, and it was suggested that it was now 'no longer economically prudent for us to produce all these components internally' (Northern Telecom 1988).

Northern's DMS SuperNode public network switch was described as having the capability to transform its public network exchanges into intelligent nodes which would give public telecommunication operators the capability to custom design and configure their own networks and services. Combined with the company's line of PBX equipment, the Meridian series, the Northern family of products was described as supporting an *open* network architecture. Northern described the future trend in network development as one in which the mixing and matching of networks and components would create a 'patchwork of partial networks' (Northern Telecom 1989).

Intelligent network Service Switching Points (SSPs) would provide the basis for interconnecting and accessing proliferating and partial networks. Northern Telecom would be a major stakeholder in the market for SSPs. However, company representatives acknowledged that there was no single perspective on the optimal technical design for the intelligent network. For Northern, the architecture simply referred to a model structured around the Bellcore concept. The main aim of the design was to unlock hardwired relationships in public switching equipment which supported routing, billing and numbering to enable the RBOCs to compete with the long-distance carriers.

Although one representative of Northern Telecom who was interviewed for this study regarded the intelligent network as a major discontinuity in network design equivalent to the shift from crossbar to digital switching,[27] other equipment designers suggested that most of the proposed services were not particularly new. Instead, the new network design would require the replacement or upgrading of existing public network equipment and so boost the future market for sales of Northern's public switching products.

Several of the Northern Telecom interviewees for this study noted that major developments in software would be needed before a truly vendor-independent intelligent network could be implemented. The company was aware that most PTOs did not want equipment manufacturers to control the functionality in their new intelligent network systems. Nevertheless, it was control of the intelligent network components that Northern believed would enable the equipment manufacturers to dictate development terms both to the PTOs and to other potential manufacturers in the computing and software industries. Northern did not intend to rely on other suppliers for the supply of the Service Control Point or the intelligent databases. The company's representatives believed that computing firms would not be able to produce reliable products for many parts of the telecommunication industry. Northern also argued that its main concern was to defend market share and that this had nothing to do with 'clever strategies for users'.[28]

Northern Telecom representatives suggested during interviews that agreement on standards and open architectures was the best way to enable manufacturers to improve the speed of design and development of products for public and private telecommunication networks. Although the company claimed it would not use proprietary standards to keep computing companies out of its market or to assist the PTOs in strengthening their hold on their markets, it acknowledged that most enhancements to existing network capabilities were taking the form of proprietary features. For Northern Telecom, the key question in the market for advanced telecommunication products was, 'Do you have an architecture that is sufficiently broad that you can gracefully implant on that architecture what the market needs – and do it earlier than anyone else can?' (Owen 1989). The company suggested that it would be some time before the agreement of common standards would cease to threaten the competitive advantage of Northern Telecom as a manufacturer, and therefore the company would opt for proprietary implementations in certain critical areas of product development.

## Regulating Network Design – The Federal Communications Commission Role

Until 1985, the FCC attempted to maintain a division between the monopoly and competitive activities of AT&T and the RBOCs by requiring that competitive (non-basic) services be offered through structurally separated businesses. Accounting practices and the costs of service supply in the competitive and monopoly segments of the companies' operations were expected to be transparent. However, the FCC abandoned its structural separation regulations when it concluded that these rules were suppressing the introduction of new advanced services.[29] AT&T had argued that structural separation requirements imposed costs in terms of the unavailability of certain services, lost economies and efficiencies, and that they prevented customers from obtaining complete telecommunication and data-processing solutions from a single vendor.

The FCC also found that the RBOCs' ability to use their monopolistic position in the local exchange market to cross-subsidize had been reduced by divestiture, the growth of competitive alternatives to their telephone services, and political and regulatory pressure at the state level to keep local telephone rates down (United States 1990a: 5749). The FCC observed that the RBOCs' ability to discriminate against those who wished to access the public network by providing inferior network connections had diminished because of industry-wide co-ordination of network interface standards. This conclusion was reached despite evidence of differentiation in the detailed designs that were being implemented by the RBOCs in their networks. The RBOCs' control over the bottleneck facilities in the local exchange plant was also believed to have diminished as a result of the threat that advanced information service competitors could bypass the local exchanges by turning to the competing network access providers (Mudd and Starkey 1992). By the late 1980s, the Idealist model which assumed the presence of widespread and effective competition was gaining ground again.

But the Strategic model still had adherents. In 1990, the United States Court of Appeals found that growth in competition in the advanced information services industry, bypass technologies and divestiture had not altered the fact that the RBOCs retained a monopoly on the bottleneck facilities in the local exchange (United States 1990a). The Court found that 'if anything, increased competition in the enhanced services market simply increases the BOCs' [Bell Operating Company] incentives to shift costs so they can engage in predatory price cutting as a means of maintaining or increasing their share of the market for enhanced services' (United States 1990a: 5759).

Despite considerable restructuring in the telecommunication marketplace the RBOCs were found to have as great an incentive in the early 1990s to engage in cross-subsidization between competitive and monopoly services as they had upon their creation in 1984 (United States 1990a: 5729). The divestiture of AT&T had not eliminated the Bell System's control over local exchange facilities; it had simply transferred control from a single national company to seven regional Bell Operating Company holding companies (United States 1990a: 5743–4).

The Court's decision argued that the RBOCs continued to control innovations in the public network which contribute to the design of intelligent services. Independent service providers still confronted the risk of being relegated to the role of imitators of the dominant network operators. With the strength of their control over access to intelligent network databases intact, the RBOCs were also in a strong position to use customer proprietary network information as a marketing tool to benefit their own information service operations. By preventing direct access to this information by competing service suppliers, the RBOCs had retained substantial market power (Federal Communications Commission 1990). The combined effects of market restructuring, multiple new entrants and technical innovation had created few, if any, of the characteristics of the Idealist model propounded by the advocates of de-regulation. The Strategic model has proven to be closer to the current reality of the US marketplace than the Idealist model. However, where once regulators had insisted upon structural separation to protect the public interest, by the late 1980s they had turned to network design criteria and Open Network Architecture standards as major new regulatory tools.

Regulation, mainly by *technical design*, has become a more predominant mode of regulation in the US telecommunication market, but the intelligent network is challenging its workability in stimulating effective competition in the market. AT&T, the RBOCs and the telecommunication manufacturers have introduced a maze of intelligent services which have increased the complexity of the public network. They have found ways of introducing tailored pricing schemes that have stretched the capacity of the regulatory commissions. Their ONA plans which enable access to the public network by competing service providers and customers have been approved in broad principle. But does this mean they have relinquished control over network access and no longer retain the market power to disadvantage unfairly their competitors or some of their other customers who wish to access the public network?

The FCC's ONA decision was intended to ensure that the RBOCs could not extend their monopoly over basic telephone services to the new generation of intelligent network services. The RBOCs were expected to unbundle and price network resources in ways that would enable competitors to repackage intelligent functionality in order to offer services to a segmented and differentiated market, but ONA was not working to the satisfaction of those who wished to gain access to the public network. This is illustrated by the complaint to the FCC by the US telephone answering service industry that, with the possible exception of Ameritech, ONA was being used to restrict competition. New entrants argued that, although the RBOCs were offering ONA within the letter of the law and new unbundled services were being made available on the same terms and conditions to competitors as to the RBOCs, many of these services were not useful as a basis upon which to build new services. In fact, many of the 'new' services were 'old' services which had been re-tariffed as Basic Service Elements. These and other new entrants argued that many of the 'new' tariffed services offered by the RBOCs were compatible with the way in which the RBOCs designed and configured their networks, rather than with the way the answering services companies and other competing service providers needed to access network resources if they were to operate profitably (*Communications Daily* 1990).

In spite of these and other expressions of concern about the RBOCs' continuing monopoly power, in September 1991, under the explicit direction of the Court of Appeals, Judge Harold Greene gave the RBOCs permission to enter the information content provision business. They would now be able to offer services such as electronic publishing, enhanced facsimile, voice response services, electronic data interchange and on-line databases in direct competition with companies which had entered the market during the late 1980s. The judge said he was removing the ban on the RBOCs' information service activities with 'considerable reluctance' since he had been instructed to do so. In his view, the RBOCs still controlled the local exchange and they retained the ability to distort the so-called competitive marketplace.

**Implications of Policy Innovations and the Network Vision**

Seen in historical perspective, many of the challenges to policy-makers and regulators created by the intelligent network architecture are not particularly new. These challenges include the need to determine whether the provision of new networks and services and the terms and conditions of their use are disadvantageous to any group of suppliers or users. The policy question continues to be whether rivalry among the market participants results in market outcomes that lead to biases in network access that can, or should, be addressed via public intervention. The difference between earlier periods in the evolution of telecommunication in the United States and the 1980s and 1990s is that companies such as AT&T and the RBOCs can no longer be certain that a failure to respond to multinational user demand in the market will permit them to maintain their market share.

A relatively small, but significant, number of large telecommunication users have realistic and cost-effective alternatives to the networks and services provided by AT&T and the RBOCs. If they do not like the services on offer they can turn to other service providers, not only for their voice and data transmission needs, but also to obtain advanced network management and information services.

Under current legal and regulatory arrangements in the United States, however, the RBOCs have both the incentive and the ability to impede competition by provisioning their advanced information services via features and functions that are technically and economically *efficient* only in the context of their own integrated networks. In an environment in which the majority of telecommunication subscribers continue to depend on RBOC network access lines to acquire telecommunication services, and even the larger customers who bypass the RBOCs still access some of their customers and suppliers through the RBOCs' local exchanges, the RBOCs have retained a strong position in the telecommunication service market.

Nevertheless, developments in the relationship between public and private networks do suggest that a new balance between the organization and control of these networks is being established. In the United States, a combination of technical change, market restructuring and regulatory intervention has created an uneasy balance between monopolistic and competitive segments of the market. The FCC has found it exceedingly difficult to regulate effectively within the context of the changes that have taken place in the US market. The regulatory tools of technical standards and criteria for open network interfaces have not succeeded in creating a market environment that meets, or approximates, the conditions of the Idealist model.

In the United States, the characteristics of the market suggested by the Strategic model are prevailing, despite the fact that ONA is premised upon the implementation of an open network architecture by AT&T, its long-distance competitors and the RBOCs. ONA has not succeeded in prising open critical points in the network or unbundling software functionality in a way that eliminates barriers to new entrants. Convergence of telecommunication and computing technologies is also in evidence only at the periphery of the public network and in support of advanced service applications. The relatively long experience of the FCC in coping with the challenge to implement rules and regulations which will promote the emergence of a competitive *level playing field* has not eliminated the considerable opportunities that AT&T and the RBOCs have retained to exercise their market power.

The network operators and the telecommunication equipment manufacturers in the United States continue to rely on proprietary specifications for the implementation of the advanced intelligent network. The process of rivalry appears to favour the large telecommunication users and to create opportunities to shift the costs of network modernization onto those customers who have the most restricted choice of services. In addition, there are signs that the intelligent network design process is providing opportunities to implement proprietary network interface standards and to seek new ways of *monopolizing* the telecommunication marketplace. This process is still very much in evidence in the United States, where competition has been favoured by policy-makers in a growing segment of the market since the 1960s. There are signs that the effects of monopolization in the US domestic market are gaining strength with the trend towards mergers among the RBOCs and the cable television operators, the Clinton administration's proposals for the provision of a core infrastructure development project and the consolidation of the facilities-based long-distance carriers. The potential for a relatively closed oligopolistic system of supply to emerge in the domestic market is strong. The forces of competition have presented themselves in accordance with the tenets of the Strategic model. The policy issue in the United States, as in Europe, is how best to ensure that public-interest goals are achieved within this evolving framework.

## The European Challenge

The intelligent network was not simply an innovative engineering design which showed promise from the point of view of its technical potential. It emerged in tandem with the RBOCs' strategies to retain existing markets and to develop new ones. The new architecture is providing a framework within which sophisticated proprietary network designs are being incorporated into the fabric of the public network infrastructure. It is being used as one of several ways to define the network access and control conditions for all suppliers and users.

If the network designers in the United States are finding new ways to control network access through technical design criteria, what are the implications for telecommunication network evolution in the European market? To assess this question we must look more closely at the process of network design in Europe. Bar and Borrus have observed that 'different communications policies directly shape economic performance by opening and foreclosing opportunities for more effective co-ordination of resource allocation and by favouring or frustrating the experimentation and learning that shape dynamic performance' in the telecommunication industry in the United States (Bar and Borrus 1989: 4). In Europe, the timing of policies aimed at introducing liberalized markets for telecommunication has been different from that in the United States. The institutional context – political and economic – also operates in a different way. The next chapter examines recent changes in the telecommunication policy environment in the European Community and in France, Germany, and the United Kingdom within the Community as well as Sweden. Emphasis is given to the role of measures that have been aimed at creating open network access conditions.

## Notes

1 The FCC's Hush-a-Phone decision, following an earlier denial and subsequent Court appeal, in 1957 allowed interconnection of this device to AT&T's network; see Federal Communications Commission 1955; Noll and Owen 1987.

2 See Federal Communications Commission 1959.

3 See Federal Communications Commission 1969b; United States 1982a; United States 1982b.

4 Some analysts focused on the failure of regulatory institutions and the legal system to respond to the new technical realities; see Horowitz 1989; Irwin 1984; Irwin and Niman 1988.

5 See United States 1980.

6 On 1 January 1984, AT&T divested itself of its twenty-two local exchange telephone companies, the Bell Operating Companies, as part of a settlement of the Government's Department of Justice 1974 antitrust suit against AT&T; see United States 1983b. The seven RBOCs are Ameritech, Bell Atlantic, BellSouth, Pacific Telesis, Southwestern Bell, NYNEX and US West. In 1980, AT&T had a total of 1,044,000 employees. With the divestiture agreement implementation in 1984, AT&T was left with 384,000 employees. The seven RBOCs accounted for 587,000 employees, for a total of 971,000. Two years later, the RBOCs had reduced their total to 547,000 and AT&T had announced a 1987 target of 290,000 (Noll and Owen 1987). The 1985 revenues of the RBOCs from regulated telecommunication services and the production of *Yellow Page* directories were: Ameritech US$2.42 billion; Bell Atlantic $2.45 billion; BellSouth $3.20 billion; NYNEX $2.59 billion; Southwestern Bell $2.20 billion; Pacific Telesis $2.26 billion; and US West $2.24 billion (Noll and Owen 1987).

7 With the implementation of divestiture, new terminology was established to describe the segments of the US telecommunication market. LATAs (Local Access and Transport Areas), POPs (Points of Presence), etc. define the boundaries of the territories where RBOCs, AT&T, and new entrants compete.

8 See Gannes 1985 and United States 1982a; United States 1982b.

9 In fact, by the time of the triennial review of the divestiture decree in 1990 even the staff of the antitrust division of the Department of Justice advocated the release of the RBOCs from most restraints on their activities as well as the release of restraints upon AT&T; see United States 1987; Selwyn and Montgomery 1987.

10 Trebing has argued that it was in AT&T's interest to use its dominant position to exclude new entrants through pricing policy, interconnect restrictions, use of facilities restrictions, terminal attachment restrictions, procurement policies, pre-emption of use of the radio frequency spectrum and contesting applications for new entry by firms before regulatory agencies; Trebing and Melody 1969: 311.

11 In September 1991, Judge Harold Greene, who is responsible for ensuring that the terms

and conditions of the Modification of Final Judgment are met, announced that the RBOCs could enter the information content business. However, the judge did not allow his decision to take effect until appeals by the newspaper publishers and information service providers had been heard. This position was overruled by the Court of Appeals.

12 The cost burden could be borne by the companies' customers and/or by the shareholders with substantial implications for the future structure and level of prices of network access and interconnection and for telecommunication service supply.

13 See Federal Communications Commission 1969a; Federal Communications Commission 1971; Federal Communications Commission 1979; Federal Communications Commission 1981; Federal Communications Commission 1987; Federal Communications Commission 1988; United States 1983a; United States 1990b.

14 Interviews with representatives of NYNEX and US West representatives.

15 And as long as measures of price elasticity continued to indicate that consumers would be willing to pay a higher price for existing local services.

16 By the end of 1988, Bell Atlantic intended to convert to CCSS7 in 150 of its New Jersey Bell total of 213 switches; forty of the Bell Pennsylvania and Diamond State switches, and forty of the Chesapeake and Potomac switches. Bell Atlantic announced plans in 1989. BellSouth and US West announced trials for 1990 and 1991, respectively.

17 Projections in 1991 called for CLASS services to be available to some 85 percent of American residential subscribers by 1995. In 1991 about 5.8 percent of equipped households had taken up some CLASS services, and this figure was expected to grow to 23 percent by 1995, representing a growth in revenues from US$2 billion to $230 billion (Mallinson 1992).

18 ONA tariffs were to replace Feature Group access arrangements for the RBOCs and the enhanced service providers with Basic Service Elements (BSE) and Unbundled Basic Access (BSA) arrangements. The BSEs are the local switching features that enhanced service providers' use to offer services. BSAs are the underlying access services that support various combinations of BSEs.

19 Interviews with representatives of the RBOCs in the United States and in Europe in early 1991.

20 Advanced Freephone services or Advanced 800 services.

21 The increase in the percentage of digital switching and transmission facilities between 1980 and 1990 was as follows: Local Loop – 6 to 15 percent; Local Exchange – 20 to 40 percent; Intra-city Transmission – 36 to 90 percent; Long Distance Exchanges – 30 to 70 percent; and Inter-city Transmission – 5 to 75 percent; source: Teledimensions, San Diego, California, 1990.

22 This proprietary model is used in the AT&T network and has been adopted by British Telecom in the United Kingdom and Telefonica in Spain.

23 In the American network, the Common Channel Signalling System (CCSS) operates as follows: a call comes into an AT&T 4ESS switch, which passes destination information to one of forty network control points (NCPs), on the CCSS network. Each control point is equipped with an AT&T 3B20 processor. These have distinct databases that support intelligent network services such as freephone, credit card authorization and AT&T's Software Defined Network. The AT&T 3B20 processor examines the packet header which signifies the source of the call, its destination and the service required. If the computer has the database, it examines the information, performs the service and connects the source to its destination. When the computer does not have the database, the call is sent to the relevant NCP over a 56-kbit/s packet switched connection. Redundant paths ensure that each call reaches the correct database and these connect all AT&T's NCPs.

24 From 1890 to 1956, Northern Electric produced equipment for the Canadian market under licence from Western Electric, the supplier of 80 percent of the American equipment market. After the 1956 Consent Decree in the United States, Western Electric withdrew from joint research and development and, by 1962, had sold its interest in Northern Electric. In 1971, Bell Canada and Northern Telecom formed Bell Northern Research, a jointly owned R&D facility. In 1972, technical information from Western Electric ceased flowing freely between Western Electric and the Canadian company.

25 This is due, in major part, to its position as the main supplier to Bell Canada, the largest

Canadian telephone company. In 1990, Northern Telecom's net income rose to US$460.2 million from $376.5 million in 1989. The company had 1990 sales of US$6.8 billion.

26 Under this contract, Northern Telecom and its subsidiaries are obligated to supply Bell Canada, and prices charged to Bell Canada must be as low as those available to Northern Telecom's favoured customers. An annual audit is used to verify that prices meet the conditions of the supply contract (Waverman 1990).

27 Interview with Northern Telecom, 24 August 1990.

28 One example of the gap between product design and the requirements of customers is the Northern's launch of the Meridian Data Networking System in the United States in 1988. This combined PBX systems with an integrated front-end processor to handle protocol conversion among disparate network architectures. This venture into intelligent networking was withdrawn in October 1989. Described as an Open Systems Interconnect compatible intelligent network architecture to provide access between information users and resources under common network management, this data network was expected to interwork with disparate products.

29 See Federal Communications Commission 1983; United States 1985.

# 4

# Latecomers or Innovators? The European Policy Challenge

> Structural crises of adjustment are thus periods of experiment and search and of political debate and conflict leading ultimately to a new mode of regulation for the system.
>
> (Freeman 1988: 10)

## Restructuring the Telecommunication Policy Environment

By the beginning of the 1960s major challenges to the monopoly regime in the United States were well underway. In Europe, telecommunication markets were divided into protected national monopolistic regimes until the mid-1980s. These national telecommunication infrastructures were provided by Post, Telegraph and Telephone Administrations (PTTs). Vertical and quasi-vertical integration between these organizations and their equipment suppliers were commonplace, and the relatively closed relationships made new entry difficult. The PTTs tended to enter into long-term contracts with their favoured equipment suppliers. In instances where such arrangements were challenged, the PTTs argued that long-term linkages were required to guarantee high standards and compatibility among the components of the public network. Thus, the need for a single operator to retain the integrity and protect the quality of the public switched telephone network was promoted by the European telecommunication operators just as it had been by AT&T in the United States (Commission of the European Communities 1990b; Commission of the European Communities 1991a).

This traditional environment is being pulled asunder in the 1990s by several factors. These include technical change, the growing presence of private networks and pressures for liberalization. Because the challenge to the European national monopolies came several decades after the first inroads of competition in the United States, there has been a tendency to treat regulation and market restructuring in Europe as *latecomer* strategies. The relevance of the US experience of telecommunication liberalization for European suppliers, users and policy-makers is frequently debated in terms of whether it carries lessons which should be adopted in Europe, or rejected as inappropriate to a very different institutional culture. This chapter illustrates the degree to which some aspects of European policy and regulation are emulating developments which have occurred in the US marketplace.

Competitive entry in both network infrastructure and service markets in Europe is being permitted in an ever-growing number of countries. In response, the Public Telecommunication Operators (PTOs), as the PTTs have now come to be called, have embarked on a programme of reforms. They hope to deliver more efficient and innovative services and many plan to become regional or global service providers. They are also seeking innovative ways of protecting their markets.

There have been two major pressures for change in the European telecommunication markets. The first has stemmed from the coincidence of investment in digital transmission and switching equipment with major political and economic developments in the European Community. Improvements in the telecommunication infrastructure have become pivotal to the challenge to achieve regional coherence based upon the development of the Single European Market (Carpentier et al. 1992; Mansell et al. 1990).

The European economic and political agenda of forging a more integrated regional market has been coupled with the second main source of pressure for change. The Single European Market is being opened to competitive entry by firms based inside and outside the European Community, and the telecommunication sector has not been an exception (Mansell 1989). The internationalization of telecommunication markets, combined with the experimentation with advanced communication services which had already accrued to American-based multinationals by the mid-1980s, brought a new stimulus for reform in telecommunication which has reverberated throughout Europe. In addition, the European-owned suppliers of equipment and services, especially since 1984 and the AT&T divestiture, have begun to seek opportunities to exploit the large US market. There have been setbacks and withdrawals from the US market, but, in general, the notion that trade reciprocity will eventually create open entry conditions in the United States as well as in Europe has stimulated a drive to develop a more efficient and competitive supply community in Europe.

The European PTOs have been encouraged to open equipment procurement to a wider number of manufacturers, partly to stimulate the competitiveness of suppliers such as Siemens, Alcatel and GEC Plessey Telecommunication (GPT). The PTOs, and especially British Telecom, France Télécom and DBP Telekom, have seen both the threat and the reality of competitive entry in their national markets. Swedish Telecom has also faced new entry in its market. In some cases, the new rivals have targeted the periphery of the PTOs' markets – for example, value-added services – but in others, they have challenged the core service, the voice telephony market, which generates some 90 percent of the PTOs' revenues. None of the PTOs had seen their markets substantially eroded at the end of the 1980s, but all saw a need for repositioning in domestic or international markets (Commission of the European Communities 1990a; Commission of the European Communities 1991a).

Freeman has emphasized that 'in a period of rapid technical change, the established social and institutional framework no longer corresponds to the potential of a new techno-economic paradigm' (Freeman 1988: 10). The new mode of regulation for the European telecommunication environment is still uncertain. Nevertheless, changes in policy and regulation are intended to ensure that rivalry and the monopolization of markets do not bring advantage only to the multinational telecommunication users. They are also aimed at extending and upgrading the services available to smaller firms and to residential consumers as well as at reducing regional disparities.

## The Incentives for Transforming Telecommunication Markets

In the Europe of the 1980s, the PTOs faced a rapidly changing set of economic incentives in the marketplace. They were confronted by new entry from service providers and the long-term threat of infrastructure competition. The response of the European PTOs has been to introduce institutional structures for implementing advanced services which differ from those in the United States. Unlike some of their long-distance operator counterparts in the United States which had faced restrictions on the provision of certain advanced services following the 1984 divestiture, the European PTOs have had few constraints placed upon the services they can offer (Commission of the European Communities 1990a; Scherer 1991).[1] However, like the competitive telecommunication suppliers in the United States and the RBOCs which generally have retained a monopoly in the local exchange market, the European PTOs are struggling to define the boundaries of telecommunication supply markets. The aim is to maintain control over the terms and conditions of access to the network infrastructure.

The European Commission is shaping the policy environment in which this struggle is unfolding. By 1990, the Commission had introduced a number of Directives regarding terminal equipment competition, equipment testing and certification, rules of procurement,[2] service competition and the terms and conditions of access to public networks.[3] On several occasions by 1990, the Commission had used its powers under the Treaty establishing the European Economic Community (1957) or the Treaty of Rome as amended by the Single European Act (1987) to signal its intention to create a liberalized telecommunication environment encompassing all the member states.[4] As the press observed, 'this is really a big tussle about whether monopolies should be maintained . . . or whether they should rapidly be liberalised. Many people are using the argument about Article 90 as a smoke screen to hide opposition to deregulation' (*Financial Times* 1989a). On paper at least, the effect of some of the Directives would make it impossible for the PTOs to prohibit the interconnection of leased lines with the public switched telecommunication network by competitors who wished to provide new services which had been targeted for liberalization.

The European Commission's Directives have been couched in general terms. The tools and mechanisms needed to define the practices required of PTOs under the new liberalized regime have been left to later stages of detailed specification.[5] Nevertheless, in the late 1980s and early 1990s, the European PTOs could not expect their practices to conform to the letter of the law and to continue to enforce rules which had restricted new entry and maintained barriers to entry in the past. If barriers to entry were to be created in the new environment, they would need either to be the result of greater efficiencies and innovativeness than potential competitors, or they would need to be less than transparent to public authorities. The Idealist model suggests that the mechanisms employed in the new liberalized market would be based on fair terms and conditions for competition available to all participants in the market. The Strategic model suggests otherwise. The PTOs would be expected to seek a variety of strategies and tactics in a bid to retain their dominance in their national markets. By 1992 the one, albeit significant, area to remain legally in the PTOs' domain was the supply of voice services and the corresponding infrastructure. This residual right would provide a platform upon which the PTOs could erect new entry barriers in the face of competitive threats to both advanced and traditional service markets.

**Transforming the Telecommunication Network**

The Directive which has the greatest direct bearing on the issue of public network control and access is the *Framework Directive on Open Network Provision* (Council of the European Communities 1990b). Open Network Provision (ONP) was defined in the Green Paper on Telecommunication published by the European Commission in 1987. ONP was described as

> a set of common principles regarding the general conditions for the provision of the network infrastructure by the Telecommunications Administrations to users and competitive service providers, in particular for trans-frontier provision. Conditions of access to the network concern mainly: standards and interfaces offered for interconnections; tariff principles; and provision of frequencies. (Commission of the European Communities 1987)

This Directive aims to make access to standardized network functions and services uniformly available and to prevent abuses of the PTOs' ability to control access and use of public networks (Analysys 1991; Council of the European Communities 1990c; Scherer 1991). Open network provision was directed initially to the 'harmonisation of conditions for open and efficient access to and use of public telecommunication networks and, where applicable, public telecommunications services' (European Telecommunication Consultancy Organization 1990: 7). Ultimately, the goal was to create a level playing field for competitive services (Gilhooly 1989a). This would require the PTOs to *unbundle* the functions embedded in the public network, to introduce independent charges for bundles of network resources, and to do so on a transparent and non-discriminatory basis (Analysys 1991: 9). This Directive clearly challenged the PTOs' claim that complete control over the public network was necessary to protect the quality and security of the public network and to ensure continuing technical innovation.

Although the ONP Directive has been regarded as a way to enable private service providers to access public networks, compromises were introduced during the drafting stages. This became necessary in order to reach agreement with the member states without requiring the Commission to force through its plans using its powers under the Treaty of Rome's Article (90)3. Such action by the European Commission had been challenged when it had been used to introduce liberalization in the terminal equipment market. As a result of the negotiations, adherence to ONP became voluntary and the ONP rules are not applicable to private service suppliers.

If ONP is to become an effective means of countering the market power of the PTOs, standardized interfaces will be needed to enable the interconnection of public and private networks. The European Telecommunication Standards Institute (ETSI) was asked to take up the challenge of developing the required standards, having been the recipient of earlier mandates from the Commission (Hawkins 1992b).[6] ETSI established a Strategic Coordination Group for ONP in March 1991, and this Group was directed to develop ONP-related interfaces for leased lines, packet switched public data networks, ISDN, voice telephone services and intelligent networks.

It was clear by June 1991 that work on the network interfaces which would encourage third-party operators to offer services using intelligent network functionality would be excluded from the standardization schedule, or at least be subject to considerable delay. Instead, ETSI chose to concentrate on defining the protocols between the intelligent network Service Control Point and the Service Switching Point (Evagora 1991). The absence of a strong push to develop standards in critical areas can be seen as a move to ensure that the PTOs would retain control over the conditions under which customers could access and use public network resources. The definition of intelligent network access interface standards was not expected to be completed until 1996 at the earliest. In the meantime, interim proprietary standards are being developed and implemented by the PTOs and the equipment manufacturers.[7] The result will be the emergence of many proprietary implementations of intelligent network-based services, regardless of the European Commission's goals of establishing a basis for greater harmonization of advanced communications service offerings across the member states (European Telecommunication Consultancy Organization 1990: 60, 62).

Article 3 of the ONP Directive states that 'Open Network Provision conditions must not restrict access to public telecommunications networks or public telecommunications services *except* for reasons based on essential requirements, with the framework of Community law' (Council of the European Communities 1990b; emphasis added). Among such essential requirements are the security of network operations, maintenance of network integrity, interoperability of services in justified cases and protection of data. If these requirements can be used to justify the failure to implement open network interfaces, the PTOs will have a potent tool to create new barriers to entry to their existing markets. Such a strategy would also effectively slow entry into advanced communication and information service markets.

The American concept of Open Network Architecture (ONA) and the European Open Network Provision (ONP) share the same aim from at least one perspective (Analysys 1991: 54). Both are intended to ensure that new service providers have fair and equal access to the public telecommunication network. However, whereas ONA is targeted at the technical interfaces, ONP is concerned with groups of services and common principles. 'Put another way, ONA tackles the problem through the technology, while ONP tackles the problems through the institutions' (Analysys 1991: 54).

The ONP approach removes from the scrutiny of policy-makers and the regulatory apparatus the very network interfaces which could lead to a level playing field in the market. The fact that the specification of standards for ONP has tended to assume that technical solutions adopted for network interfaces are appropriate to a quasi-competitive, monopolistic environment is to be expected when the gestation of the concept is considered. ONP was defined initially by the European Commission's Analysis and Forecasting Group (GAP) which reported to the powerful Senior Officials' Group on Telecommunication (SOG-T). Members were drawn almost exclusively from the European PTOs. The European Council of Telecommunication Users Associations (ECTUA) has claimed that their members were not involved sufficiently in the processes of framing the Directive. Other user groups have also argued that they were less than fully involved; for example, the European Association of Information Services, the International Telecommunication Users Group and the International Chamber of Commerce.

Developed in close association with the ONP Directive was the Services Directive (Commission of the European Communities 1990a). This Directive came into effect in 1990 and abolished exclusive or special rights for the PTOs over all services except voice telephony. Private Service Operators (PSOs) were eligible to offer value-added services. From 1 January 1993 basic data transmission would be opened to competition, except in special circumstances which would apply to PTOs in the Less Favoured Regions of the European Community. Once again, one of the aims of the Services Directive was to limit the power of the PTOs to control access to the public network.

The PTOs had managed to retain their monopoly on voice telephony – defined as real time speech – and many continued to prohibit the simple resale of leased lines. In 1992, the Commission published a plan for the further liberalization of the market.[8] Although early drafts called for a complete opening of the voice telephone market and the public network infrastructure to competition, in October the Commission was calling for consideration of four options.

Option 1 Freeze the liberalization process . . . and maintain the status quo;
Option 2 Introduce extensive regulation of both tariffs and investments at the Community level in order to overcome the bottlenecks and in particular the surcharge on intra-Community tariffs;
Option 3 Liberalize all voice telephony – i.e. international (inside and outside the Community) and national calls;
Option 4 An intermediate option of opening to competition voice telephony between the member states.

The Commission also called for the further development of open network provision and its application to voice telephone services to ensure Community-wide interconnection and universal access to the public network at reasonable conditions and quality (Commission of the European Communities 1992a: 25,30).

By late 1991, only Germany, the Netherlands, Denmark and the United Kingdom had taken significant steps to implement ONP. To date, a limited number of ONP offerings have been defined and several interface standards have been published. Equipment suppliers have expressed fears that the PTOs 'which still play a dominant role in the standards-making process – will seek to turn the harmonisation process to their own advantage' (Analysys 1991: xiii). Harmonization of the conditions required under the ONP Directive across the Community member states is not equivalent to the standardization process. For example, ETSI could develop publicly agreed standards with regard to interfaces which *may* be implemented by manufacturers. Harmonization calls for the introduction of legislation by the member states to ensure that specified European standards are *actually* implemented.

## European Problems of Regulatory Design

Compliance with most of the Commission's Directives for telecommunication was not progressing well by mid-1991. Infringement proceedings were launched by the Commission against Spain, Italy, Portugal, Greece and Ireland for non-compliance with the Services Directive. The Commission also charged all member states except the United Kingdom and the Netherlands with violations of the ONP Framework Directive. Since May 1991 most member states have countered that they are in compliance, but Figure 4.1 suggests that the reality is considerably different.

The ONP Directive introduced the concept of 'equivalent transmission capacity' (ETC) to enable the PTOs to introduce competitive service offerings on transmission capacity which differs from that available to competitors. It was suggested in the trade press that

> ETC would create the opportunity for the telecommunication organisations to charge their own competitive activities lower prices for the same transmission

| | BEL | GER | DEN | FRA | GRE | ITA | IRE | LUX | NETH | Port | UK | Switz |
|---|---|---|---|---|---|---|---|---|---|---|---|---|
| Terminals Directive implemented? | Partial | ✓ | ✓ | ✓ | Partial | Partial | ✓ | ✓ | ✓ | ✓ | ✓ | ✓ |
| Any exclusion? | Yes | No | No | No | Yes | Yes | No | No | No | Yes | Yes | Yes |
| **Services Directive** | | | | | | | | | | | | |
| Special and exclusive rights withdrawn? | ✓ | ✓ | ✓ | ✓ | ✓ | No | ✓ | No | ✓ | ✓ | ✓ | ✓ |
| Regulatory and operational powers separated? | No | ✓ | ✓ | ✓ | No | ✓ | ✓ | No | ✓ | ✓ | ✓ | No |
| Resale over PSDN? | No | ✓ | By deadline | By deadline | No | By deadline | No | ✓ | ✓ | ✓ | ✓ | ✓ |
| **ONP Framework** | | | | | | | | | | | | |
| **Draft Leased-Lines Directive** One-stop shopping? | ✓ | Under negot'n | Under negot'n | N/A | N/A | Under negot'n | ✓ | Under negot'n | Under negot'n | Under negot'n | ✓ | ✓ |
| Basic set of leased lines available? | ✓ | ✓ | ✓ | ✓ | N/A | ✓ | ✓ | ✓ | ✓ | ✓ | ✓ | ✓ |
| PSDN minimum set of offerings by 1994? | ✓ | ✓ | ✓ | N/A | N/A | N/A | ✓ | ✓ | N/A | N/A | N/A | ✓ |
| ISDN minimum set of offerings by 1994? | ✓ | N/A | ✓ | N/A | N/A | N/A | ✓ | N/A | N/A | N/A | N/A | ✓ |
| **Draft Voice Telephony Directive** Position on Draft Directive | Will conform | Will conform | Will conform | N/A | N/A | N/A | None at present | No comment | No comment | No comment | No comment | N/A |

*Note:* N/A indicates that no reply was given. No reply had been given by the Spanish authorities at the time this report went to press.
*Source:* Analysys and *Communications Week International*, 11 May 1992

Figure 4.1 *ONP implementation, European Community, 1992*

> capacity. For example, a PTO charges its competitors one price based on the fully distributed cost while it may charge its own competitive activities another price based on the incremental cost of ETC. (Gilhooly 1991c)

This observation contains warning signals for public policy-makers seeking to monitor and arbitrate an issue which has bedevilled American regulators for some twenty-five years. Judgements as to the relevant costs of public network resources, the transparency of methodologies used to calculate them, and the extent to which the results are applied without discrimination to the public network supplier's own activities and to its competitors have been the source of continuous controversy in the United States and will become increasingly important in Europe. The ETC concept has also brought the complex world of technical design and standardization to the forefront of the policy debate. Policy-makers and -regulators are faced with arguments over which technical interfaces and standards represent *equivalent capacity* and what costs should be assigned to these components of infrastructure. The resolution of these matters will affect the development of open public network access and the extent to which PTOs can implement technical market entry barriers through the standards-setting process.

The technical interface standards and the network design problems are not the only issues with which the European Commission must contend. The ONP Directive is also concerned with the tariffs charged for access to public networks and with broader pricing strategy issues. The press argued in 1989 that, despite the fact that new entry in telecommunication service markets could present the spectre of cream-skimming on the most lucrative routes served by the PTOs, the solution was 'not to restrict competition but to encourage it, so as to stimulate the PTTs to rationalise their tariffs and improve their efficiency' (*Financial Times* 1989b). 'Cream-skimming' is a term used to describe the tendency for new entrants to provide services in areas characterized by high traffic volumes and relatively low costs, as compared to low traffic-volume, high-cost areas. It is argued, generally without recourse to empirical studies of actual costs, that the former are the lower cost routes served by the incumbent operators.

Pricing controversies have not been restricted to the public switched network. Leased line tariffs have been high too, and, with the exception of the United Kingdom where British Telecom's market had seen the growth of private networks, this alternative remained costly in most of the other member states. In February 1989, British Telecom, the Deutsche Bundespost (DBP), and the Dutch and Danish PTOs opposed the new tariffs for leased lines which had been suggested by CEPT.[9] The European Commission found that CEPT proposals for international leased line pricing were anti-competitive. A consultancy study had shown that European telephone companies were charging between four and twenty times the equivalent prices for international leased lines (Rogerson 1989).[10] Observers of the structure of intra-European telephone tariffs have concluded that prices for services which average two-and-one-half times those of the highest national long-distance calls cannot be justified on any basis of the cost of supply (Touche Ross Management Consultants 1991). These pricing strategies were extremely unlikely to be deemed to be *cost-related* as required under the Commission's new policy regime.

A European Commission investigation of PTO pricing strategies was prompted by complaints by the European Council of Telecommunication Users Associations and the Bank of America,[11] and by April 1990, the Commission had launched an inquiry into international telephone prices within the Community. The FCC in the United States had argued that overcharging for international telecommunication services was pervasive in Europe. The problem of cost-related pricing emerged again in the spring of 1991 in the negotiation of the prices for links connecting competing digital cellular telephone network operators. The press observed that a solution might be found in the implementation of cost models, and argued, for example, that 'achieving cost-based tariffs depends largely on whether the operator succeeds in setting up a high-performance system for calculating costs, both in terms of software and hardware' (Blau and Schenker 1991). The crucial issue as to which costs ought to be included in the prices for public network access paid by competing operators in Europe did not appear to surface publicly at this time.

If the European Commission was moving slowly to force major changes in European telecommunication price structures, the International Telecommunication Union was appearing to bow to the pressures for change. The International Telecommunication Union's CCITT had set up a working party to determine the costs that should be reflected in the tariffs for leased circuits, although this work was forecast to continue for about ten years. The CCITT also proposed a revised Recommendation that would permit companies to resell spare capacity using the public telecommunication network, share telecommunication circuits and, with some limitations, interconnect private networks with the public networks.[12] The beneficiaries were to be resellers of simple bandwidth, providers of managed end-to-end international circuits, suppliers of hybrid private and public international service packages, and so on (Lynch 1991a). If this recommendation had taken effect immediately in the European Community, it would have seriously challenged and undermined the PTOs' pricing structures, which bore little relationship to the underlying costs of international and domestic service supply. Since changes at the international level would be slow to take effect, the PTOs in Europe did not act immediately to restructure their tariffs.

### Uneven Network Development Problems

The vision of a ubiquitous advanced intelligent network is often presented as the logical outcome of the full deployment of digital technologies throughout the telecommunication infrastructure. The conclusion will be the emergence of a malleable network with transparent boundaries between

Table 4.1 *Network evolution in the United States*

| Date | Service components |
|---|---|
| 1900s | Cord Switchboards (using party lines) |
| 1920s | Limited Local Dialling (operators still required to place many metropolitan area and most rural area calls) |
| 1940s | Introduction of National Direct Distance Dialling (most manual switchboards eliminated, use of party lines all but gone except in rural areas, Touch Tone introduced as premium service option) |
| 1970s | Widespread Deployment of Analogue Stored Program Control Electronic Switching Systems, replacing electromechanical central offices (full mechanization of toll billing, limited introduction of central office based 'custom calling services') |
| 1980s | General Availability of International Direct Distance Dialling (extensive deployment of digital carrier on interoffice and interexchange trunks, 'Equal Access' to interexchange carriers, basic and 'enhanced' 911 service, extensive use of public 'voice' network for data communications) |
| 1990s | Full Deployment of Common Channel Signalling at the end office level (introduction of many new software-based network features; introduction of digital plant for business and residential subscriber access lines; adoption of Touch Tone as the 'standard' offering; deployment of new 'Open Network Architecture' interconnection and network access arrangements; introduction of limited ISDN at the subscriber level; implementation of TDD/voice relay systems) |

*Source:* Adapted from Lee Selwyn (1990) 'Network modernisation: an evolving view of basic telecommunications service'. in R. Gabel (1992)

public and private services. This is the vision of adherents to the Idealist model, who then argue the following. During the transition from the old, established technical regime, based upon inflexible analogue equipment and hierarchical network architectures, the PTOs must be constrained from preventing entry by companies better able to meet large (and often small) business and consumer demands for advanced services. Policies and regulatory intervention are needed to ensure that monopolists are restrained and public services are developed that can be justified on grounds of efficiency and public service. However, once the transition is over, regulatory intervention in the market should be superseded by the superiority of the market as a mechanism for stimulating efficiency, innovation and diversity in services. This is the Idealist model in full swing.

This excessively simplistic model takes no account of the dynamics of technical change and the uneven development of public and private network infrastructures. The reality that is reflected in the Strategic model is a transition to strategic oligopoly that replicates, and may even exaggerate, disparities in the accessibility of the public network. This is the case, even in the face of declining equipment costs and innovation in technical facilities.

This point is illustrated by trends in the evolution of the European telecommunication network which are highlighted below.

*Public Network Modernization*

The determinants of the uneven pattern of development of the public network can be considered in the light of two questions. The first is the question of who should pay for the costs of telecommunication modernization. The second is whether the public switched network should support all the leading-edge service applications. To answer these questions it is necessary to consider the history of the telecommunication development cycle. This is illustrated for the United States in Table 4.1. A similar pattern of investment has been followed in other countries, although the timing and ubiquity of network modernization has differed.

Each phase in the process of telecommunication modernization has required the PTOs to address technical problems with regard to network organization, transmission, switching, signalling and so on. Inevitably, these problems are linked with issues of cost recovery and pricing policy (see Table 4.2). At the root of pricing controversies are concerns that arise with technical change and shifts in patterns of network use. Very often telecommunication service prices have not been modified by the PTOs to take account of changes in the way the telecommunication system is designed and how costs are incurred to provide services. Prices have also not been restructured to reflect the changes in the services (and customers) that benefit *most* from the deployment of costly capital equipment.[13]

Implicit in the recent upgrading of the public network is the idea that more intensive utilization of the network for new data and image services, especially those required by large users, will permit costs to be spread over a broader range of services and customers. But in an environment characterized by strategic oligopoly there are economic incentives that argue in favour of the recovery of increased costs caused by the introduction of these new services from customers who are not the main users of services. This is but one of the central questions relating to the integration of local and long-distance networks and the present conversion to intelligent networks.

The second question relates to whether the public switched telecommunication network should support all the leading-edge service applications. The historical evidence is that this problem is one of timing. Rarely have co-ordinated initiatives been taken to equip the whole public switched network with the most innovative technical solutions to changing communication requirements. For example, the digitalization of public switched networks in Europe has shown considerable variation in the timing and targeting of investment in different parts of the network.[14] Even where massive investment in public infrastructure has been introduced in advance of a clear demand structure for service applications – for example, the French videotex network – the upgraded public network which has resulted has not met the full requirements of multinational corporations. These firms

Table 4.2 *Technical change and network modernization: the key constraints*

| | |
|---|---|
| Problem | What technical problems do the provision of a particular service create? |
| Technical solution | What equipment or organizational changes are made to overcome those problems?<br>Network organization:<br>Who talks to whom and how is a connection established?<br>Transmission:<br>What is the medium over which the communications are sent?<br>Switching:<br>How are messages from one subscriber to another subscriber routed?<br>Signalling:<br>How is the status of the system indicated for purposes of controlling the flow of traffic?<br>Numbering:<br>How are messages addressed so that they can get to their destination?<br>Management accounting:<br>How are accounts identified and transactions recorded for billing purposes? |
| Cost implication | How much does it cost to effect the change and deploy the necessary equipment? |
| Regulatory response | How do regulators treat the increased cost (or revenue) that flows from the solution? |
| Price impact | How do the charges for various services reflect regulatory or policy decisions? |

*Source:* R. Gabel (1992)

have also taken advantage of the private network alternatives to public switched infrastructure capacity and they have introduced considerable intelligence into the periphery of the public network.

### *The Emergence of Private Networks*

Private networks using capacity leased from the PTOs have a growing presence in Europe and the telecommunication bypass phenomenon is growing as well (OECD 1992: 83). In those markets where competitive infrastructure supply is permitted – for example, in the United Kingdom, radio, satellite and cable, and in Germany, satellite and radio – the trend towards diverse forms of self-supply supporting voice, data and image communication is gaining strength.

In the United States the availability of higher capacity – for instance, 1.5 Mbit/s equivalent – T1 links is already pervasive.[15] European initiatives to stimulate the availability of higher capacity networks operating at speeds in excess of 2 Mbit/s have, so far, resulted in gradual improvements, but several consortia have faltered in their attempts to upgrade trans-European telecommunication links. For example, the European Broadband Interconnection Trial (EBIT) initiated by sixteen CEPT members, soon experienced technical, managerial and tariff problems and the European Commission withdrew its funding.

In September 1991, British Telecom and France Télécom withdrew from this initiative just as plans were being formulated to commercialize a pan-European network to offer higher bandwidths on demand. Another initiative, METRAN (Managed European Transmission Network) involved European PTOs. Advanced digital technologies were to be used to support switched and leased networks operating at high capacity.[16] However, services were unlikely to be available until the 1995–98 period. HERMES, the European railways' initiative,[17] depends for its future on the lifting of regulatory constraints on the intra-European, trans-border supply of switched and private line services. Plans include fibre-based supply of services operating at 64 kbit/s and 2 Mbit/s at prices that are competitive with those available in the United States. The proponents of these projects intend to use high capacity fibre-based systems to challenge the dominance of the European PTOs.

Another challenge to the PTOs has come from the liberalization of the satellite market in Europe. In 1990, European satellites were carrying only 2 to 3 percent of intra-European traffic. Private VSAT (Very Small Aperture Terminal) receive-only networks have existed only since the 1988 European Commission terminal equipment Directive liberalized ownership and operation of one way satellite terminals.[18] In early 1989, the Department of Trade and Industry in the United Kingdom awarded seven Specialized Satellite Service Operators (SSSOs) licences and restricted operators initially to providing national point-to-multipoint service without interconnection with the public switched network. This move was designed to protect British Telecom and Mercury from bypass of their public networks. In November 1989, SSSOs were allowed to uplink signals from Britain, to broadcast them to Europe, and to downlink signals anywhere in the United Kingdom. Two-way VSATs have also been liberalized in Germany, and they can be interconnected with the public switched network. In 1992, the Commission introduced its Satellite Directive challenging both the PTOs and Eutelsat, the main space segment operator, to liberalize the market further.[19]

Finally, the mobile radio market in Europe has grown significantly. Although these operators have concentrated on the voice telephony market, they have challenged the PTOs in national markets where competitors also have been licensed.[20] Cable television licensees in Britain have shown that, at least in the short term while they compete on price and quality of service, there is considerable demand for their telephone services.[21]

It is clear that the potential exists for substantial restructuring of the European telecommunication market. From the perspective of both the Idealist and the Strategic models, the policy question is whether these initiatives will face barriers to growth as a result of the practices of the PTOs. If they do, will these be competed away as more advanced technologies diffuse through the public network, or will the outcome be the emergence of a strategic oligopoly that meets the requirements of an even smaller 'club' of users than the PTO monopolies of the past were able to serve?

The large telecommunication suppliers and users are in a strong position to allocate economic resources to the development and use of advanced telecommunication technologies. Their requirements are sophisticated, and they can often sustain the risk of failure of a costly investment programme in an innovative technique. For smaller businesses and consumers, the risk is greater and their resources are smaller. An important role for public policy is to ensure that the unevenness of public and private network development does not become exacerbated as the structure of the marketplace changes. The policy and regulatory environments in the United Kingdom, France, Germany and Sweden have shaped entry conditions in the telecommunication market in different ways according to perceptions of who should pay for the modernization of the public switched network and who the main beneficiaries should be. To make such judgements the policy regimes in these countries have encouraged different balances between public service and efficiency objectives, and they have relied on the market to substantially different degrees.

## National Policy Initiatives in Telecommunication

### *Liberalization in the United Kingdom*

The 1969 Post Office Act replaced the public telecommunication and postal administration in Britain with a public corporation and achieved the administrative separation of post and telephone services. The telephone monopoly, British Telecommunications, remained intact until the beginning of the 1980s.[22] In 1981, a new Telecommunications Act was implemented. This Act provided for the establishment of a public telecommunication corporation. It also permitted the licensing of competing network operators and liberalized the terminal equipment market.[23] The Act allowed licensing of value-added services provided using leased lines under the responsibility of the Department of Industry. Mercury Communications Ltd, then a consortium composed of Cable & Wireless, British Petroleum and Barclays Bank, was awarded an interim licence in 1982 and the *duopoly* in British telecommunication was born.[24]

The Value Added Network Services General Licence of 1982 provided a boundary for competitive services that excluded real-time voice and data communication from competitive supply.[25] It also called for case-by-case decision-making on whether new services should be treated as *basic* or *value added*. In each case a judgement would need to be made as to whether messages transmitted over the network were being acted upon, either by being stored before being forwarded, by being changed in format, code,

Table 4.3 *Mercury's financial performance, 1988–90*

| Measure | 1988 (£m) | 1989 (£m) | 1990 (£m) |
|---|---|---|---|
| Gross revenue | 64 | 170 | 351 |
| Payments to other operators | 16 | 55 | 158 |
| Net revenues | 48 | 115 | 193 |
| Depreciation | 20 | 27 | 54 |
| Trading profit | −11 | 18 | 49 |

*Note:* Based on year end 31 March.
*Source:* Fintech (1990)

protocol or content other than to facilitate conveyance to the destination, or by being forwarded to more than one person.

In 1984, the British Telecommunications Act resulted in privatization, and the Office of Telecommunications (Oftel) was charged with negotiating the difficult path between the ideal of competition and the imperfections of the actual marketplace.[26] In the same year, Mercury's licence was extended to allow it to compete with British Telecom across a wide range of services, including international voice telephony. Most of the major cities in the United Kingdom were connected to Mercury's figure-of-eight network by 1986, and, in that year, Mercury began offering switched services in addition to dedicated leased line services. Mercury's financial performance from 1988 to 1990 is shown in Table 4.3. Mercury's trading profit of £49 million in 1990 compares with British Telecom's £3.2 billion in the same year. Mercury has remained heavily dependent on British Telecom for the termination of its services and for the supply of leased capacity to connect segments of its trunk network.[27]

British Telecom's licence included some specific constraints on the company's freedom to price its services (Tahim 1990). In 1988, both British Telecom and Mercury applied to Oftel for more freedom in setting prices for value-added and data services. British Telecom requested waivers from obligations to publish prices for these services to improve its competitive strength against smaller rivals. In granting a waiver, Oftel required that the companies would have to make their accounts available for inspection.

In 1989, British Telecom asked Oftel to allow it to raise prices for Mercury's connections with its network. The company argued that Mercury should pay a commercial rate. Behind the issue of the appropriate level of prices charged for interconnection of competitors with British Telecom's public network was the issue of subsidies, their magnitude, structure and the beneficiaries. For example, in 1989, Oftel noted that at some stage Mercury might have to pay a share of the costs incurred by British Telecom to subsidize loss-making rural telephone services.

Another facet of the pricing issue concerned tariffs for international telephone calls which were believed to be generating substantial profits. British Telecom argued that this matter was not in its hands since national governments were responsible for establishing the system of cost and revenue-sharing among connecting international public telecommunication operators. They were therefore responsible for the timing of any reductions in the prices of these international services. British Telecom's Chairman suggested that 'British Telecom is, we believe, one of the few international telecommunications operators that would be happy to see the accounting rates come down. But these rates are largely determined by national governments, not by independent operators' (Vallance 1990b).[28] The company claimed that long-distance calls were subsidizing local calls in an amount well over £1 billion a year.[29]

The British Telecom position was as described as follows:

> I agree, too, with the need to define and quantify the cross subsidy with care. It currently comes in two forms. The first and by far the most significant (more than £1 billion a year) derives from the tariff distortion between call charges and exchange line rentals. Cost related tariffs, which you [Oftel] have been advocating in the context of international calls, could eliminate this first form of cross-subsidy. The second derives from conditions in our operating licence which require us to provide certain services, irrespective of their underlying economics, such as free 999 calls, uneconomic payphones and rural services, and the universal service obligation. (Vallance 1990a)

Furthermore:

> British Telecom's current tariff structure is an historical hangover. It was put together years ago during the long dead era of the old General Post Office when some cross-subsidy was normal since telephone services for both residential and business customers were in the same government accounting 'pot'. But in today's increasingly competitive and customer-focused climate, cross-subsidy is an anachronism that distorts the market and inhibits competition. (British Telecom 1990)

Nevertheless, the Director General of Telecommunication initially took the view that 'accounting estimates of cross-subsidies should be regarded with suspicion: they are almost certainly invalid' (Dixon 1990b). He noted that it was impossible to allocate costs between different types of telephone service because, for example, a telephone call cannot be originated without a telephone exchange line and a telephone exchange line alone would be of no use unless it were connected to other parts of the network so that it could be terminated.

The question of equal access to the public infrastructure was raised in the context of the Government's review of the British Telecom–Mercury Duopoly policy in 1990.[30] Implementation of equal access by British Telecom's competitors to its public network was estimated at a cost of £300 million to modify the public network exchanges. By 1991, the Director General of Telecommunications and British Telecom's Chairman were disputing the costs and prices of public network access and service use, openly, in the national press. British Telecom's Chairman brought the weight of the academic community to bear on his side stating that the company 'has advanced compelling arguments, backed up by leading economists on both sides of the Atlantic, which demonstrate that [price] rebalancing and economically sound interconnect charges are prerequisites to genuine competition in United Kingdom telecommunications' (Vallance 1991).

Mercury claimed that it would face £189 million a year in extra costs if an access fee for using British Telecom's local network were to be introduced, but British Telecom responded that

> the sections of the market which are currently underpriced are, in the main, local residential services and particularly the provision and maintenance of residential exchange lines. The result is that the emergence of a choice of suppliers to the ordinary, domestic consumer is being inhibited by the pricing regime Sir Bryan [Oftel] insists upon. . . . The bottom line on all this is that successive Secretaries of State for Trade and Industry, and Sir Bryan himself, have consistently fudged the issue of re-balancing our charges for rentals and calls. For example, earlier this year, Sir Bryan and Peter Lilley [Department of Trade and Industry] came up with the idea that BT and all other operators should pay for the imbalance through an 'access deficit contribution'. Earlier this month they effectively abandoned the idea, or at least postponed it *sine die*. (Vallance 1991)

In mid-1991, Oftel decided to eliminate the access charge payments to British Telecom until competitors had reduced the incumbent's market share to 85 percent. These decisions were taken amid fears that restricting British Telecom's right to levy access charges would jeopardize the Government's ability to raise £11 billion from the sale of the second tranche of British Telecom shares.

Oftel has faced the difficult task of reconciling the incompatible policy objectives embodied in the 1984 Telecommunications Act. For example, the Director General of Telecommunications must secure such telecommunication as satisfies all reasonable demands, providing that the supplier is able to finance these services; promote the interests of consumers, purchasers and all other users in the United Kingdom; promote effective competition in the supply of telecommunication; secure efficiency and economy on the part of the suppliers; promote research and development in the United Kingdom; and enable the United Kingdom telecommunication producers to compete effectively within and outside the United Kingdom.

British Telecom's licence has been criticized for its failure to include sufficient safeguards against anti-competitive behaviour (Beesley et al. 1987; Gist 1990). By the end of 1991, British Telecom still claimed some 95 percent of the total British telecommunication market. Although there had been the occasional rumour of major structural change, no radical changes of the magnitude experienced in the US divestiture of AT&T had been witnessed.[31] British Telecom is also subject to very few restrictions on the services it offers.[32] The company is required to provide separate accounts for some of its lines of business – for example, terminal equipment sales – and must show that cross-subsidies are not unfairly disadvantaging competitors.

It may not use information gathered in its capacity as a network operator to increase its sales or to discriminate in favour of its own customers.

In the early 1990s British Telecom's public statements echoed those expected of a competitor within the Idealist model. British Telecom's Chairman saw the post-Duopoly environment from 1991 as a return to the web of regulatory interference and control from which privatization was intended to allow it to escape. But one participant in the liberalization of the UK market suggested that the Government had not gone far enough in creating the conditions for effective competition, let alone for the operation of the Idealist model.

> It is very difficult for outsiders to break into this vertically integrated package. BT does not encourage interconnect, or equal access, and it prefers to continue selling its service as whole 'bundles'. It is essentially impossible to buy part of a BT 'bundle' without buying the rest. This *frustrates competition and denies choice to end users*. Sadly, the duopoly policy review has accepted the continuation of these 'bundles'. I think it is a mistake. (Ellison 1991: 3)

In the event, an uneasy alliance between the adherents to the Idealist and the Strategic market models has been struck. The Idealist model could be used as justification by those who felt it was appropriate to unleash British Telecom from restrictions on its pricing practices. The Strategic model was pressed into service and reflected in measures intended to restrain British Telecom, and any future dominant firm, from anti-competitive behaviour. The history of liberalization has shown that, whatever its political motivations, the shape of the market is substantially influenced by interaction among the major institutions of control: the telecommunication suppliers, the companies using the public telecommunication infrastructure and the regulatory apparatus. Furthermore, the beneficiaries of change have been decided by the negotiated outcomes of price, cost and network access debates.

### *The Adjustment Process in France*

The French Government took command of the postal and telegraph operations as early as 1879, and the telecommunication network was developed in a monopolistic framework. Although initially private concessions were permitted, these had all been nationalized by 1889 (Dang N'guyen 1988). By 1923, a special budget for the Ministry of Post, Telegraph and Telephone had been created, and, in 1944, the Centre National d'Etudes des Télécommunications (CNET) was established to undertake R&D and to develop standards for telecommunication equipment.[33] In 1968, the Direction Générale des Télécommunications (DGT) was created in order to draw a distinction between the Ministry (Ministère des Postes et Télécommunications) responsible for telecommunication and the operation of the telecommunication system.

Since the early 1970s, telecommunication operations have been restructured to achieve an increasing separation of postal and telecommunication services.[34] A new telecommunication law was to have been introduced in 1987 to increase the responsiveness of telecommunication network supply to the changing European and international environment. This was delayed until 1990 because of the strong reaction of the trade unions against far-reaching changes in the public organization and the desire on the part of the socialist government to move slowly towards significant change in the telecommunication field.[35] Discretionary decrees were introduced throughout the mid- to late-1980s in an effort to stem growing concerns about the limited opportunities for the supply of competitive services in the French market.

In 1986, a Mission à la Réglementation Générale was created within the Ministry for Post and Telecommunication to encourage change in the structure and operation of telecommunication (and the postal service). It was from approximately this year that controversies concerning the future orientation of France Télécom, the name adopted by the DGT in early 1988, began to be debated.

The Mission à la Réglementation prepared a draft Law for Telecommunication in 1987. The law defined basic service as the transmission of signals that are not processed, and included in this category telephony, telex, circuit and packet switched data transmission and leased lines. This approach incorporated more services in the basic category than were then under consideration by the European Commission for its reserved (monopoly) service category. The law also proposed two separate categories for networks: the general network operated by France Télécom which would meet social and political obligations, and competing networks which might be authorized to complement the general network.

Apart from telephone services, other information services could be offered in competition with France Télécom if they were provided using leased lines. No detailed rules existed with respect to such services until a decree with respect to specialized services open to third parties was introduced in September 1987.[36] This decree created a category of value-added services which could be provided by competitors if they were authorized or declared. To provide such services, the cost of the transmission component of a service could not exceed 15 percent of the total cost and no voice traffic could be carried. Leased lines would continue to be priced on a flat-rate basis, and although provisions for surcharges were introduced they were never implemented.[37]

Issues of interconnection and the definition of network interfaces would be defined by the Minister to guarantee the integrity of the public network. Provisions were made to enable France Télécom to levy charges for the use of the telecommunication network, for recovering the costs of adapting switching capabilities to provide for access, and for contributions to the obligations imposed by the state on France Télécom. The debate on the introduction of public network access charges was thus introduced at a relatively early stage in France.

Tensions were evident in the struggle to control the regulatory aspects of telecommunication. These emerged in 1989 with the adoption of a decree which substituted the Direction de la Réglementation Générale (DRG) for the Mission à la Réglementation, still within the Ministry of Post and Telecommunication, but outside the scope of France Télécom itself (Coriat 1989). In the interim, between the drafting of the telecommunication law in 1987 and the passage of the new law in 1990, a Commission was established to debate the future of the telecommunication sector. This debate included customers, industrial participants and employees of France Télécom.[38] It addressed the appropriate role of the state in telecommunication; the market for postal and telecommunication services; and the impact of institutional changes on ministry personnel.[39] The findings of the Prévôt Commission were released in August 1989. They called for the separation of post and telecommunication services and the transformation of France Télécom into a more commercial operation. No details as to how this should be achieved were given at this time. However, the strongest pressure for change in France appears to have been generated by external factors, such as the European Commission and the requirements of multinational firms.

With the implementation of a new legal regime in 1990, France Télécom became a corporate body under public law with a legal identity separate from the state, though it continues to be owned by the state. The new law created the France Télécom Group which includes France Télécom and COGECOM.[40] The structure of the France Télécom Group is shown in Figure 4.2. France Télécom is now under the supervision of the Minister for Post and Telecommunication and subject to the oversight of the DRG.

The provisions of the Law on Telecommunication Regulation (LRT) of 1990 were incorporated into the Post and Telecommunication code, giving responsibility to the Minister of Telecommunications for the regulation and supervision of all telecommunication activities. Article L-33-1 of the Post and Telecommunications Code reserves 'the exclusive right to set up networks available to the general public (switched telephone network, telex network, etc.) to France Télécom. . . . The "independent networks", including private or joint networks, require a Ministerial authorisation. Certain small or low-bit rate networks can be set up without prior authorisation' (France Télécom 1990: 7). As a result of these changes, varying degrees of competition have been introduced in the French telecommunication market.[41] The new law attempts to create a boundary between the provision of the public network and certain other services and competitive services.[42] The Ministry may grant or refuse authorization for services depending on whether they meet a public need and are compatible with the activities of the public operator, France Télécom. Some observers have argued that it is really only in the area of 'services including the transmission or routing of signals or a combination of these functions using telecommunications processing' that there has been the beginnings of liberalization (Dupuis-Toubol 1991: 14). The law introduced the concept of bearer services, which include simple data transmission and exclude data-processing. These can be provided by competitive suppliers if they are

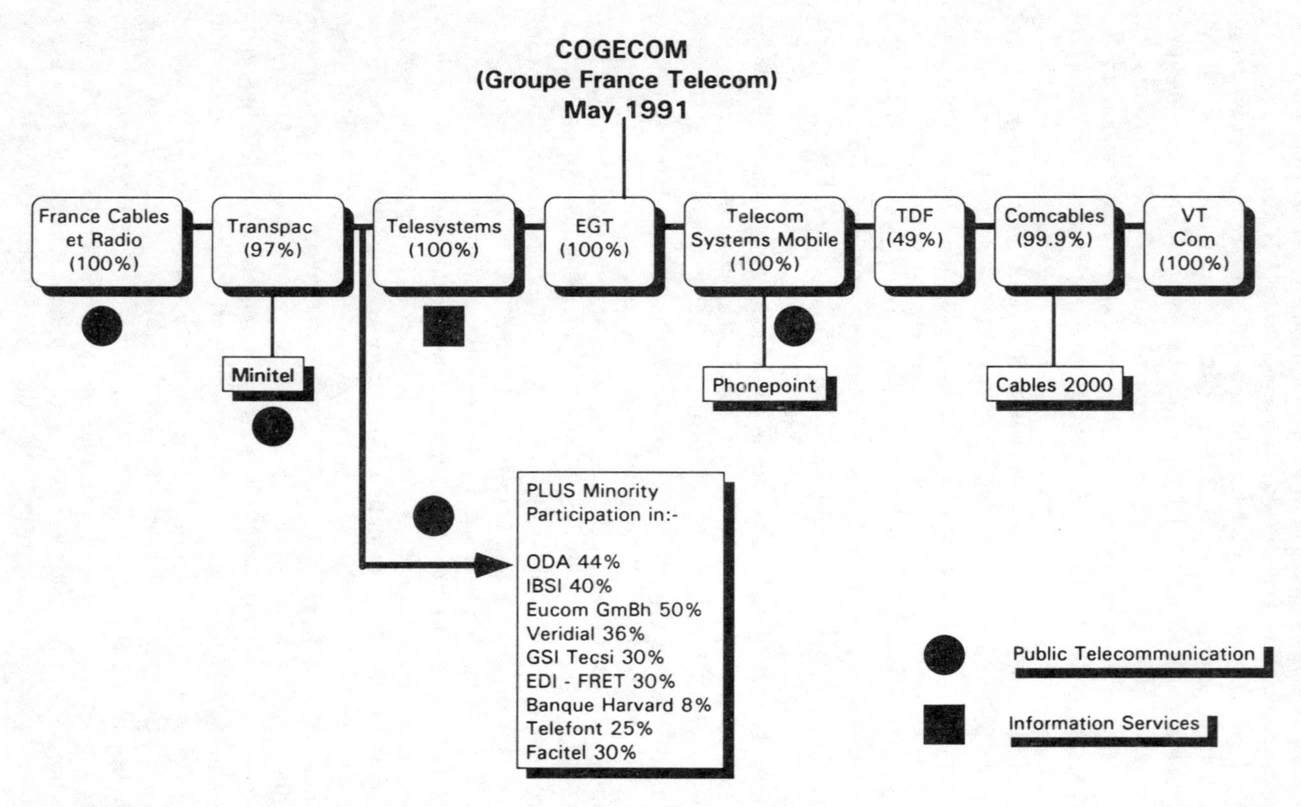


Figure 4.2 *COGECOM (Groupe France Télécom), May 1991 (COGECOM, 1991)*

authorized. The Minister must determine whether a service 'does not hinder France Télécom in the successful accomplishment of the public service obligations assigned to it and the constraints relating to tariffs and geographic coverage'(Dupuis-Toubol 1991: 14).

The new laws created a sphere of legal monopoly for France Télécom where only a *de facto* monopoly had existed before. The old Code for Telecommunication required only that the establishment of telecommunication services and networks be subject to authorization by the Minister of Post and Telecommunication.[43] However, a stronger protection of its monopolistic status has not prevented the Télécom Group from suggesting that with its new public operator status and the socio-economic environment, France Télécom must more than ever be ready to listen to its customers (France Télécom 1990: 9).

The DRG defines its mission as being to devise clear and transparent rules which will allow market participants to compete fairly. For example:

> Effective regulation is doubtless based on clear, transparent and objective rules, to which the economic players must refer to provide and use the products and services concerned by this regulation. These rules must propose a stable framework allowing these players to make good choices and provide good arbitration, especially in the long term, since it is true that the communication activities in which the post and telecommunications sectors participate take on strategic value in modern society. *Good regulation must also clearly identify public service missions and the resources needed to accomplish them.* (Ministère des Postes et Télécommunications 1992: 71; emphasis added)

The concern to ensure that public service objectives remain enshrined in the activities of France Télécom appears to have survived in spite of pressures to give primacy to competitiveness and strategic concerns. There is a perception that the bulwarks of monopoly can be maintained in the provision of the fixed terrestrial infrastructure and bearer services even in the face of pressures to achieve greater efficiencies in the provision of competitive services.[44] To achieve the goal of balancing competition and public service objectives, network design and standardization of the public network infrastructure will play an important role. For example, as the DRG has observed, 'it is necessary to maintain access to a broad uniform technical platform, bearing a diversified supply of advanced services' (Ministère des Postes et Télécommunications 1992: 73). In April 1993, with the election of a new Government, the press reported that plans for the privatization of France Télécom were being considered, thus bringing issues of the balance between public service and international strategic goals forcefully to the forefront of the political agenda.

While the regulatory regime has been undergoing change, France Télécom has not been idle. Unlike British Telecom, which orientated its network modernization activities towards the public voice telecommunication network, in France much greater attention has been given to the public data network. The largest data network in the United Kingdom had 3,500 customers in 1991, whereas France Télécom's Transpac had 5.5

million users (Abrahams 1991).[45] The linkages between the commercial strategies of the Transpac subsidiary and the main operations of France Télécom are considerable. As one observer has commented, 'it is impossible for Transpac to decide on its own about its business strategy'.[46]

In the domestic market, France Télécom has sought to ensure that it retains the business of its largest customers and it has worked to build confidence in its ability to manage networks. The company has argued that customized corporate networks combining Transfix (high-speed fixed digital links), Numéris (ISDN), and the Transpac (packet switched data) network and the range of advanced equipment provide 'complete, coherent and structured corporate networks' (France Télécom 1990).

The company's subsidiaries have access to the public network services, and there is a considerable blurring between the public sector and commercial or competitive activities. These relationships are far from straightforward (Pospischil 1988: 7). For example, France Télécom has pursued bilateral alliances to counter challenges from its competitors. In 1990 France Cables et Radio purchased an 80 percent equity stake in Cylix, an American satellite operator, and Télésystèmes purchased holdings in computer and engineering companies in Germany and Italy. Agreements have been made to supply unified service points, the France Télécom answer to multinational user pressure to provide single points of contact for business customer services.[47]

Throughout the changes in the France Télécom structure, the company's research organization, CNET, has remained under its jurisdiction. CNET has strengthened its competence in computing and software development as well as in telecommunication. The importance of this expertise has been recognized explicitly by CNET's President. In comparing the structure and competitiveness of telecommunication in the United States and France, he has pointed to the differences in the ways in which design and technical expertise evolved in the two countries in the 1970s and 1980s.

> The expertise in the United States in the area of multiplexing was a result of its deregulation and tariffs policy which has fostered the development of high-speed networks that make use of multiplexing. Meanwhile, in Europe and in France, in particular, we have preferred to focus on more sophisticated networks and services, such as packet switching or the videotex network. (Bidal and Mangin 1991: 31)

The intelligent network is one of the key areas in which CNET has been developing specifications.

### *Transformations in Germany*

Historically, the Deutsche Bundespost (DBP) has been the responsibility of the Ministry of Posts and Telecommunications, and its operations, regulation and political functions have been combined.[48] From the 1920s, the DBP was a *special fund* separated from the general budget and no subsidies were allocated to it. This resulted in a certain ambivalence. The organization was financially independent but it was also politically dependent upon government (Haid and Mueller 1988; Neumann 1986). The political character of telecommunication issues required that tariff changes be discussed in the German Bundestag. The Telecommunications Installation Act gave the DBP a right to a monopoly in telecommunication, but the DBP did not engage directly in manufacturing, choosing instead to co-operate with its main German equipment suppliers.

The legal background for the West German telecommunication environment contributed substantially to the strength of the monopoly which existed until 1990. First, the DBP is mentioned in the Basic Law, and its organization is fixed unless a two-thirds majority vote in the Bundestag is achieved. This was not considered likely to change but in mid-1992 political debate began to address the privatization option for German telecommunication. Until this time manufacturers, unions and the major political parties had been opposed to any major change along these lines.

In comparison with the experience of British Telecom, some observers suggested that, during the 1980s, the DBP had retained a comparatively good public image. The public network operator had begun to introduce special concessions to meet the telecommunication requirements especially of the banking sector. Other segments of the business community also were believed to have received special treatment with respect to regulations and tariffs in order to minimize complaints about the unresponsiveness of their monopoly supplier.[49]

A Government Commission was established under the chairmanship of Eberhard Witte in 1985 to consider telecommunication liberalization measures. In its recommendations in 1987 the Commission favoured the retention of a monopoly for the public network infrastructure and for voice telephone services and the liberalization of other segments of the market. Structural and accounting separations were also suggested to curtail internal cross-subsidies (Witte 1987). However, as Schmidt has observed, the Ministry of Posts and Telecommunications had also been formulating proposals for liberalization (Schmidt 1991). The Ministry timed its proposals to coincide with the release of the Witte Report. The Ministry sought ways to extend and protect the DBP's monopoly on voice telephony and even to extend its protected market to new real-time voice advanced services. The changes that ultimately were introduced reflected a compromise between these two sets of proposals.

The new arrangements retained the monopoly on the public telecommunication network, and the *Postverfassungsgesetz* (the Deutsche Bundespost Constitution Act) came into force in July 1989. The definition of the network monopoly was restricted to the transmission element of voice telecommunication service supply. This was a concession to private service providers who argued that they required the use of their own switching capacity to provide services. According to the Act's provisions, the Federal state was given the exclusive right to set up and operate transmission lines, including the associated network terminations and the network monopoly, and to establish and operate radio installations. Other service providers were given the right to provide telecommunication services to third parties over leased lines or switched connections which would be provided by DBP Telekom.[50] Voice telephone service remained in the monopoly segment of the market as did services involving certain innovative features.[51] Equality of access to leased circuits was guaranteed, as was interconnection with the public telecommunication network. Although no details were provided in the initial agreements, Open Network Provision was to be implemented.

A mandatory service category was introduced to cover public service obligations imposed upon DBP Telekom; for example, nation-wide coverage, equality of access, tariff averaging and so on. These were to be subsidized if necessary by the monopoly services.[52] Finally, DBP Telekom was permitted to enter into any competitive market.

As the changes in the German telecommunication market have taken hold, DBP Telekom's responses to the threat of competition have become apparent in the company's pricing strategies. The liberalization of the telecommunication market has been accompanied by pressures to move prices closer to *costs*. The old DBP had introduced usage-sensitive tariffs for certain categories of leased line services in the 1980s and the outcry from large users was considerable (Neumann and von Weizsaecker 1982).[53] As a result of the pricing structure in place during the 1970s and 1980s, most services offered by private suppliers used the public switched network (Heuermann 1987). Closed user groups had begun to share capacity and to take advantage of the use of leased lines, but stringent restrictions were used to define eligibility for participation in closed user groups (von Weizsaecker 1987). Pressures for new tariff policies increased when the Bank of America National Trust and Savings Association lodged a formal complaint with the European Commission, charging that the national monopolist was breaching competition law by penalizing users of private networks (Schenker 1990).

Helmut Ricke, Chairman of the Management Board of the DBP, described 1990 as an historic year. The company became a separate public corporate entity responsible for providing services in a new competitive environment. The new entity, DBP Telekom, was to be commercial 'in most respects'. The company was expected to become more outward-looking and its official comments echoed this orientation. For example:

> In the long term, Telekom will endeavour to integrate the products it offers into a global range of services in the international arena. Important objectives are uniform standards, interlinking internationally available networks, achieving the position of a European hub for network operation, and harmonisation of tariffs and services. (DBP Telekom 1991b: 42)

Until the initial telecommunication reform process was completed in mid-1989, the DBP was regarded as one of the least liberalized of the PTOs in western Europe (*Financial Times* 1989d). During the 1980s, public officials had categorically rejected the privatization of the DBP as a way of stimulating efficiency and positioning the public network operator in the world market. However, by 1991, the newly reorganized DBP Telekom was making references to the future possibility of privatization in spite of the need for a change in the basic law (DBP Telekom 1991b: 17).

The first Telekom Annual Report following the unification of East and West Germany in 1990 and the implementation of liberalization plans described the organization as 'a market-oriented and customer-oriented enterprise operating in an increasingly competitive environment' (DBP Telekom 1991b). The East German PTO, Deutsche Post, was merged with DBP Telekom in May 1990 (Neumann 1990), and DBP Telekom took responsibility for services and operations in the five East German states in October of that year.[54]

The DBP Telekom of the 1990s is coping with the construction of a new telecommunication infrastructure in the eastern Federal states. The east German modernization plan, Telekom 2000, is expected to cost some DM 55 billion over seven years, and the estimates are constantly being revised upwards.[55] In August 1990, the Ministry of Posts and Telecommunications announced the conditional suspension of the DBP Telekom monopoly on voice traffic in order to stimulate the establishment of satellite links between east and west Germany. By March 1991, only one company, Preussen Elektra of west Germany, had successfully applied for a satellite voice traffic licence, but many others had received licences for non-voice satellite links. A further relaxation of restrictions on competition in the voice telecommunication service market resulted in the introduction of six-year licences for voice service competitors (Goodhart 1991a). Thus, DBP Telekom, like its counterparts in the United Kingdom and France, was struggling with widespread challenges to its monopoly power.

Nevertheless, the German operator claimed that it intended to remain one of the top three network operators in the global market (DBP Telekom 1991b). To achieve this objective, the company argued that it 'must have the same freedom to act in all . . . areas as its competitors in the private sector' (DBP Telekom 1991b). The company aimed to globalize its business and establish strategic alliances with other network operators. Globalization was described as an effective means to achieve a new division of labour in the provision of telecommunication services, to reduce costs and to create new synergies in the competencies of existing and new service and equipment providers (Ricke 1991: 5).

### *Competitive Inroads in Sweden*

The first few decades of telephone development in Sweden were characterized by intense competition. By the early 1900s, the telecommunication administration, Televerket (known as Swedish Telecom in the 1980s) commanded a major share of the market. Public utilities such as railways and electricity and defence departments were permitted to operate their own networks. The Swedish parliament deliberated for ten years as to whether a statutory monopoly should be created. The *de facto* monopoly which

Swedish Telecom had succeeded in building was sanctioned without the introduction of protective legislation.

The absence of a *de jure* monopoly has been a contributing factor to the perception that telecommunication policy in Sweden is markedly different from that in other western European countries. In fact, one representative of the company has seen greater similarities in the United States than in other European countries (Thorngren 1990). Swedish Telecom has always operated separately from the postal service and, despite the fact that it has been government-owned, its management has claimed substantial independence from the government. Decision-making power has been vested in the Board of Swedish Telecom, and the Ministry of Communications has defined general telecommunication policy objectives.[56] However, the company has not been entirely free from governmental oversight. It has been subject to policies concerning the general economic framework; for example, the fair trade ombudsman, the National Price and Competition Authority, the ombudsman for consumer protection and the National Audit Agency (Nordling 1990).

In 1980, the Swedish parliament took a decision which led to the gradual liberalization of the telecommunication equipment market. Unlike other PTOs in Europe by this time, Swedish Telecom had installed plugs and sockets:

> in almost every room in every building, for decades there is no way to make any distinction between the 'first, second, third' and so on, telephone. The customers simply buy the sets, plug them in, and, if needed, also install any further inside wiring on a do-it-yourself basis. (Thorngren 1990: 3)

The 1980 decision by the Swedish parliament also led to the establishment of Teleinvest AB, a commercial company within the Televerket Group. This organizational structure was intended to facilitate access to financial markets and to develop the commercial standing upon which to enter joint ventures. Teleinvest was required to account separately for its activities, and this encouraged the Swedes to look to the United States for analogous experiences of the telecommunication liberalization process – for example, structural separation requirements (Nordling 1990). The Swedish steps along the road towards liberalization are shown in Table 4.4.

Competition in the supply of the public network services began in 1981 with the establishment of COMVIK, a competing cellular network operator.[57] In 1988, COMVIK proposed to interconnect its international satellite service, COMVIK Skyport, to the Swedish Telecom public network. The entry of COMVIK raised the question of network interconnection, and provided a glimpse of the degree to which Swedish Telecom had retained its bargaining power. An interconnect agreement between the two companies was reached but not before objections had been raised by Swedish Telecom. From this point, Swedish Telecom faced the effects of the full liberalization of receive-only VSATs and two-way satellite transmissions, although these still required radio frequency usage permits.

Table 4.4 *Liberalization in Sweden, 1979–90*

| Year | Event |
|---|---|
| 1979 | First call for Customer Premises Equipment (CPE) liberalization. |
| 1980 | Swedish parliament, gradual liberalization (establish Teleinvest AB). |
| 1981 | Teleinvest AB established. |
| 1981 | COMVIK – cellular competition, interconnect PSTN. |
| 1981 | Nordic Mobile Telephone System (NMT). |
| 1982 | Initial CPE liberalization. |
| 1985 | Further CPE liberalization. |
| 1988 | Swedish parliament, telecommunications policy guidelines. |
| 1988 | COMVIK, Satellite Telepost. |
| 1988 | Notelsat, VSAT field trials. |
| 1988 | Scandinavian Telecoms Services (STS) with Nordic countries. |
| 1989 | Final CPE liberalization. |
| 1989 | National Telecommunications Council (STN) equipment approval. |
| 1989 | Ministry of Transport and Communications, study group on ONP. |
| 1990 | Swedish Telecom International AB. |

*Sources:* Various

Competition also had been felt with the increasing penetration of cable television which, by mid-1989, served about 40 percent of the potential market. The Swedish railway and electricity organizations also posed a potential competitive threat to Swedish Telecom as did Tele X, the satellite operated by the Swedish Space Corporation. All these organizations controlled transmission capacity which could be used to provide data and voice services in the domestic and international markets.

Bohlin and Granstrand (1989: 16) have been sceptical as to the degree to which these organizations posed a serious competitive threat to Swedish Telecom. They have suggested that the close links between the electricity, rail and telecommunication industries in Sweden are sufficient to inhibit collaboration between the electricity and rail companies and potential telecommunication service suppliers. They also have argued that there was little chance of these utilities entering directly into the market in the late 1980s.

In 1988, the Swedish parliament proposed that telecommunication regulatory tasks should be separated from Swedish Telecom's operations, in line with the general trend towards independent public telecommunication operation that was being promoted by the European Commission. A National Telecommunication Council, the Statens Telenamnd (STN), was established to consider questions of interconnection by competing service providers. Initially, the STN was charged with registering equipment attached to the network, and it was unclear whether it would take on the controversial issues surrounding interconnection of networks provided by competing operators. The new body was not given responsibility for consumer protection and it was not intended to accumulate a staff that was large enough to take on an interventionist regulatory stance.[58]

The proposal called for Swedish Telecom to continue to manage the problem of interconnection without intervention in the initial stages of negotiation by the state but with the aim of ensuring that all competitors received equal treatment. Disputes were to be referred to a fair trade ombudsman, and it was recognized that separate Swedish Telecom personnel should handle technical information from the interconnecting operators. Swedish Telecom argued that these measures would provide the basis for ensuring that telecommunication would be subject to the same laws and procedures as any other business. For example, the company suggested that 'Telecom is an area that is changing too quickly to be imprisoned in too detailed regulation, which could also put market development in a strait-jacket' (Nordling 1990: 7). The company was a strong advocate of the Idealist model of telecommunication development.

Swedish Telecom argued that value-added services were flourishing in the domestic market and that competing providers have 'established themselves freely, without any barriers to entry' (Nordling 1990: 5). But the growth of the value-added services industry was slower than expected and Swedish Telecom was relatively slow to diversify into the advanced communication services market. For example:

> The Swedish TA [Telecommunication Administration] has diversified into VANS [Value Added Network Services] but has not made heavy investments in them. The Swedish TA has instead sought collaboration and joint ventures with large users and VANS operators as its strategy has been to seek collaboration instead of competition, with the aim of increasing the total network flow. However, if the VANS industry were to start competing with the TA's role as network supplier, the Swedish TA would probably become more aggressive. (Bohlin and Granstrand 1989: 16)

In 1989, Swedish Telecom introduced the category of voice operator for competitors, with a substantial volume of voice services using the public switched telecommunication network. The treatment of these operators was described rather ambiguously as being 'based on the international situation, using the existing international calculations of cost elements for pricing national extension' (Nordling 1990: 5).

The company also appeared to recognize the legitimacy of the Strategic model. To ensure that competing service suppliers and network operators interconnecting with Swedish Telecom's switched telephone network contributed to the costs of maintaining and upgrading the public network infrastructure, Swedish Telecom proposed that a charge should be levied upon competitors sufficient to generate 'a satisfactory return on invested capital in the switched telephone network and stimulate continued commitment to utilising modern technology and maintaining sufficient capacity' (Televerket 1990: 15). The company recognized that 'some sort of regulatory agency would be necessary to mediate disputes between Swedish Telecom and other network operators involving conditions for interconnection' (Televerket 1990: 16).

Faced with British Telecom/Tymnet, AT&T International, US West, and

Sprint, GEIS Co., IBM and the sale of 40 percent of the COMVIK Skyport to Cable & Wireless, Swedish Telecom saw the need to reposition itself in the global market on the same footing as its international competitors. Some of the company's rivals are shown in Figure 4.3.

An early initiative in this direction was a response to the requirements of large business customers. Swedish Telecom entered a joint venture with the other Nordic PTOs called Scandinavian Telecommunication Services AB (STS). STS was intended to offer business users one-stop shopping services, network management and hybrid networks. In 1987 bilateral agreements were negotiated with the PTOs in Hong Kong, Australia, Japan, Singapore, Canada and the United States. But Europe proved much more difficult when the other Nordic PTOs began to consider the implications of the venture and to slow down the introduction of services. In March 1990, STS was abandoned.[59]

In its place, Swedish Telecom International (STI) AB, a subsidiary of Teleinvest, was created. In 1990, STI expected to become a top international operator through strategic partnerships to provide end-to-end service to large customers. STI believed that its pricing strategy would be essential to an aggressively growing business. The main challenge was to gain access to external markets. One representative of Swedish Telecom described this as an *Octopus* problem. For example, if all the telecommunication links were to go to Sweden like the arms of an octopus most of the revenues would remain in Sweden. In STI's view, the PTOs were building the networks but it was their competitors that were benefiting. STI also believed that customers had to be convinced of the need to use new services. The challenge was to build a 'network in a network' which would provide access to business customers.[60] Swedish Telecom was learning to respond strategically and to support competitive measures. The question was whether it could continue to do so.

By late 1990, the threat of competition had stimulated Swedish Telecom to become more aggressive and to call for its own privatization by changing its status from that of a Crown Corporation to a company in which 45 percent of its shares could be held by private investors. The company stated that unless changes were made quickly 'the unique Swedish public-service corporation will be unable to compete in the integrated European market of the 1990s' (Televerket 1990). The proposal advocated that the Swedish Telecom Group should be established as a limited company, Svenska Tele AB, which would incorporate Teleinvest and Swedish Telecom. As the company's Director General put it, 'we must follow our customers out into Europe and the world' (Taylor 1990). The opening in 1992 of a joint venture with PTT Netherlands, called Unisource, provided a clear sign of how this relatively small PTO would seek to secure its future in Europe and in the international market (PTT Telecom 1991).

The Swedish Government plans the passage of the first telecommunications law in 1993 which will codify a licensing system and give Swedish Telecom status as a state-owned corporation. The National Telecom

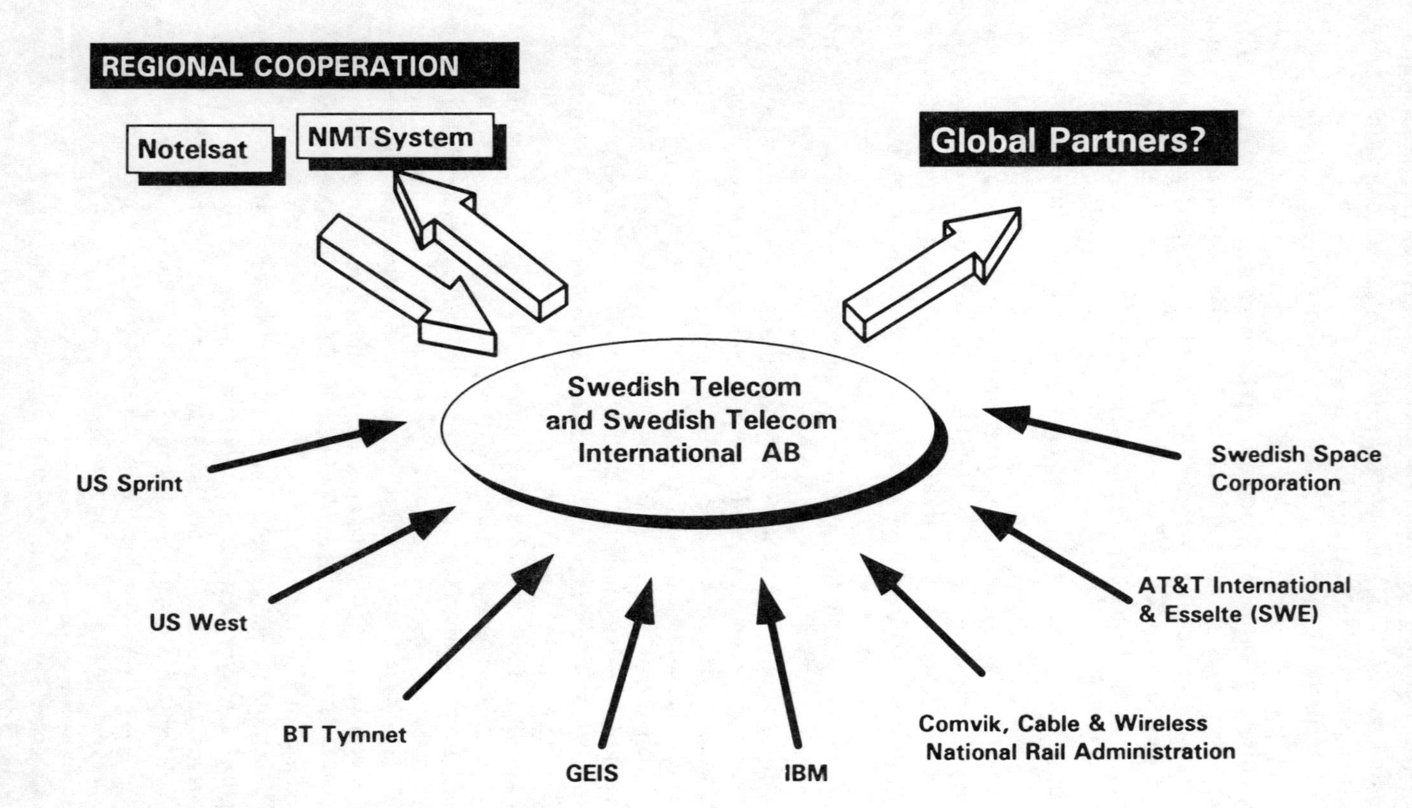


Figure 4.3 *The competitive challenge in Sweden*

Agency will provide the necessary regulation as will the competition authorities, but neither the procedures nor the strength of the Government's commitment to the development of competition in the domestic market have become clear.

**Strategic Telecommunication Policy**

The telecommunication policy regimes in Europe are developing in an environment where the PTOs and their competitors are introducing advanced intelligent services using different suppliers, technologies and standards. Network investment is permitting a variety of new options for accessing the enhanced public networks.

Most services (for example, freephone, credit card authorization, televoting and shopping, network management and virtual private networks) do not require that the intelligent network architecture be standardized according to any single model. Nevertheless, to achieve an environment in which open access to public networks can flourish in the way envisaged by the Idealist model, certain network resources must be made available by the PTOs. If these are priced as separate commodities, they can be bundled together in innovative ways to provide the basis for new (or marginally differentiated) services which can be offered by competitors. The key is the standardization of the network interfaces between users and PTOs and the prices charged for use of network functionality.[61]

The significance of network standardization and design concepts for the battle of control over revenues accruing to network and service operators as well as sales revenues for the equipment manufacturers is highlighted in a study on intelligent networks and ONP: 'The real "added value" in IN [intelligent network] services will come from the management side. For instance, service management for VPN [Virtual Private Network] is of utmost importance. Therefore, IN should provide as much flexibility in defining service management as in defining service logic to handle real time calls' (European Telecommunication Consultancy Organization 1990: 42).

The companies operating in an intelligent network environment will not be able to avoid the question of network interface standards. Interface standards could help to stimulate vendor independence enabling the PTOs to choose from a wide variety of hardware and software products. Standardized interfaces could facilitate the interconnection of services provided by different PTOs in one another's markets or enable a PTO in one country to access the network resources of another. However, most significantly from the perspective of the PTOs and their competitors, interface standards can be used by a service provider as a basis upon which to access the public network and to configure new services.

This in itself is not a threat. Nor is it simply a question of implementing optimal technical solutions to protect the security and integrity of public and private networks. As the authors of a report on the progress of ONP implementation have observed, 'we think it is necessary that the infrastructure supports some protection mechanisms (policing, statistics, auditing, tariff policy) in order to protect: the network from the Service Providers; the Service Providers from other Service Providers' (European Telecommunication Consultancy Organization 1990: 42).

The protective mechanisms that can be introduced into the network interfaces can be designed to accomplish technical goals, and they can be used to open or close markets. As Hawkins has concluded, technical decisions inevitably embrace a host of non-technical political and economic considerations (Hawkins 1992b).

In general terms, the PTO perspective is captured by the following quotation which equates standardization of certain network interfaces with a dangerous threat to network integrity:

> This solution provides perfect equality of access to the Service Providers and the PTOs. It yields average performance from the response time point of view as there is some signalling network overhead between the SCP [Service Control Point] and the network resources. In fact, it may be possible to enhance performance by defining more powerful functions (that would be more numerous than less powerful functions). Quality of Service is difficult to achieve due to the complexity of the interface, and to the fact that the Service Provider has to worry about network resource failures. Such drawbacks would not be a problem for big Service Providers but would rule out smaller Service Providers. . . . From the security point of view, the use of the interface is somewhat *dangerous* for the network integrity, and because it is impossible to monitor everything in the network. (European Telecommunication Consultancy Organisation 1990: 47–8; emphasis added)

The stated objective is to protect the integrity of the public network. There may be significant technical security problems which arise with the complexity of the public network and the pressures to open access to intelligent network resources. By giving priority to the maintenance of network integrity, however, the PTOs are able to buttress their monopolistic control over the public infrastructure by closing, or at least reducing, the accessibility of intelligent features embedded in the public telecommunication infrastructure.

At the national level, the Idealist model is much in evidence in the public statements of the PTOs and in many of those of public officials. For example, in the United Kingdom the argument is orientated towards the competitive benefits that should flow from the success of British Telecom in international markets. This should be complemented by a flowering of competition in the domestic telecommunication service market. In France, there is a selective international strategy and a reorientation of France Télécom's activities toward the business domestic telecommunication market. DBP Telekom's interest in the global and western European market has been diverted by its investment in east Germany, but competition in both infrastructure and service supply is becoming an increasingly favoured way of meeting public policy objectives. In Sweden, Swedish Telecom's alliances point to the company's growing orientation to the international marketplace.

In the following chapters, the design of intelligent network in the United Kingdom, France, Germany and Sweden is examined. The central issue is whether the trend is towards the opening of public network access in a way that conforms with the assumptions of the Idealist model. If this is the case, there could be an argument for the withdrawal of most forms of market regulation. The vision of a global *network of networks* encompassing all communication and information requirements assumes that this is the trajectory of technical innovation.

However, evidence that the opening of network interfaces between the public network and the facilities of competing service providers is encountering resistance from dominant actors, as would be expected by the Strategic model, suggests that there are a number of alternative trajectories for the evolution of global networks. The liberalization policies and the formal regulatory apparatus in the United Kingdom, France, Germany and Sweden are relatively immature as compared with those in the United States. In this sense, European countries are *latecomers* in developing the institutions and powers that will be needed to counter the monopolistic power of the traditional PTOs. In the European Community, there is as yet no regional regulatory authority that could undertake this responsibility.

However, rather than passively adopting the American approach, it should be recalled that the regulatory regime in the United States has been less than effective in coping with the Strategic model of telecommunication development. It has not succeeded in countering tendencies towards network closure, and it has not prised open the public telecommunication infrastructure provided by the RBOCs or by AT&T, at least not to the satisfaction of many competing suppliers. European policy-makers will need to learn from the fact that, in the United States, regulation has not been instrumental in promoting the emergence of the idealists' fully competitive marketplace where all barriers to entry are substantially reduced.

As European policy-makers confront the rivalry that is emerging in Europe, there are opportunities to create policy and regulatory frameworks that can more successfully counter the biases in the market that are envisaged by the Strategic model of telecommunication development. In this sense, European policy-makers have an opportunity to promote institutional innovation. The questions are whether they will do so and whether the capacity for policy and regulatory innovation can be sustained in the longer term.

## Notes

1 Most PTOs in Europe are permitted to offer all services on a competitive basis. Constraints apply to the use of specific technologies; e.g. British Telecom is not currently permitted to terminate its services to customers using radio-based technologies.

2 By the end of 1988, the European Commission had announced a Directive for the public procurement of telecommunication suppliers and work contracts. The Directive covered all bids of US$230,000 or more on all equipment – from mainframe computers and network

software to other equipment. It stated that the PTOs could reject any bid if the value of at least half the products and services came from outside Europe. Even if a bid met the 50 percent requirement, bids by European companies 'shall be preferred' as long as they are deemed to be equivalent and not more than 3 percent higher in price. These provisions were part of a Directive which covered utilities providing water, energy, transport and telecommunication. This was extended to telecommunication by a Council Directive (Council of the European Communities 1990a), and has been the source of disputes between Community and American trade negotiators since coming into effect in January 1993.

3 See Council of the European Communities 1986. By 6 November 1992, only the United Kingdom, Denmark and France had introduced the required national legislation. Germany and the Netherlands had introduced interim measures and none of the European Telecommunication Standards Institute (ETSI) Common Technical Regulations (CTR) had been completed for equipment testing (Commission of the European Communities 1990a). In January 1993, the CTRs still did not exist. They are scheduled to come into effect during this year as a result of a new Directive. At present, ETSI standards can become NETs (Norme, Européenne de Télécommunications) which are mandatory. The CTR programme is intended to replace, or at least to restructure, the NET programme.

4 Article 90(3) of the Treaty of Rome states that 'The Commission shall ensure the application of the provisions of this Article and shall, where necessary, address appropriate directives or decisions to Member States.' Article 100a states that,

> By way of derogation from Article 100 and save where otherwise provided in this Treaty, the following provisions shall apply for the achievement of the objectives set out in Article 8a. The Council shall, acting by a qualified majority on a proposal from the Commission in cooperation with the European Parliament and after consulting the Economic and Social Committee, adopt the measures for the approximation of the provisions laid down by law, regulation or administrative action in Member States which have as their objective the establishment and functioning of the internal market.

In July 1988, the French Government challenged the European Commission's use of Article 90(3) on the grounds that the usual approvals by the European Parliament and the Council of Ministers had been bypassed. With respect to the Terminal Equipment Directive, in 1991, the European Court of Justice found in the Commission's favour (European Court of Justice 1991).

5 For example, a Proposal for a Council Directive on the mutual recognition of licences and other national authorizations to operate telecommunications services, including the establishment of a Single Community Telecommunications Licence and the setting up of a Community Telecommunications Committee (CTC), was only presented by the Commission in July 1992; see Commission of the European Communities 1992c.

6 Mandates or requests to develop standards in order to ensure that the harmonization objectives incorporated in the Commission's Directives and related documents have been given to ETSI by the Commission with respect to ISDN, GSM (Global System for Mobile Communication) and several other technologies.

7 By December 1992, ETSI had completed two baseline documents: *Draft Technical Report: Intelligent Network Framework; Draft Technical Report: Guidelines for Standards*. Progress had been made on network management in an intelligent network environment, and service management and creation standards are to be addressed from 1993 to 1997.

8 Under Article A10 of the Services Directive and Article A8 of the ONP Directive or Framework Directive (Commission of the European Communities 1990a; Council of the European Communities 1990b).

9 Conférence Européen des Postes et Télécommunications. France Télécom and eight other PTOs intended to adopt a new two-tier pricing strategy for leased lines. One tier would apply to all customers using telephone-type circuits with normal quality for voice and/or data without access to the public network. The other tier would apply to telephone-type circuits connected to the public network and/or those carrying third-party traffic. These customers would pay an access charge amounting to 30 percent more than the price paid by those not carrying third-party traffic or not connected to the public network.

10 The discrepancy was generally much larger for long-distance circuits of 200 km than for short distance circuits of 2 km. In Sweden and the United Kingdom, prices were found to be roughly in line with costs although longer-distance circuits were subsidizing shorter distance circuits (Rogerson 1989).

11 A ruling overturned the CEPT Recommendation on the General Principles for the Lease of International Telecommunications Circuits and the Establishment of Private International Networks, April 1982. The Commission had considered the possibility of an exemption to the ruling under Article 85(3) in order to encourage harmonized tariffs.

12 Although the Recommendation was agreed in 1992, there is no detailed annex specifying how costs are to be established. Several countries will not remove their reservations to the Recommendation until the details are clarified; personal communication with delegate to the CCITT working party, October 1992.

13 This analysis does not extend to related issues of implementing advanced services that generate network usage information which can be used in ways that run counter to privacy protection legislation.

14 The percentage of network digitalization in different segments of the network in the countries of main concern in this study as of 1990 was as follows for digital exchange lines and trunk lines, respectively: France – 70.0, 74.0; west Germany – 12.0, 50.0; Sweden – 56.0, 75.0.0; United Kingdom (BT) – 45.0, N/A; (Mercury) – 100.0, 100.0 (OECD 1992).

15 Even in the 1970s when resale of private-line capacity had been permitted in the United States by the FCC, some 1,700 applicants had filed to compete in various segments of the long-distance market. While not all would lease capacity from AT&T, the demand for T1 links grew significantly from this early period. A survey in 1991 suggested that European annual growth of private networks for voice service was at a rate of 8.7 percent, while data networks were growing at a rate of about 31.7 percent. The Yankee Group reported growth as follows: voice all services – 8.7 percent; voice outsourced – 5.5 percent; voice insourced – 19.6 percent; data all services – 31.7 percent; data outsourced – 27.3 percent; data insourced – 39.7 percent (*Communications Week International* 1991c). In the United States in 1990, it was reported that private network users were pressuring network management vendors to provide facilities that had not come on the market (International Resource Development 1990). There was a very low percentage of American users with ISDN plans and low penetration. In late 1990, ISDN connections represented less than 0.5 percent of all exchange lines.

16 Synchronous Digital Hierarchy and Digital Access Cross Connect supporting circuit switched and leased lines using Asynchronous Transfer Mode.

17 Includes NYNEX International, Sprint International, TeleColumbus AG, Tractebel SA, Daimler-Benz AG and Compagnie de Suez. Racal Electronics plc withdrew in 1991. The Racal subsidiary, Racal Network Services Ltd, was to concentrate on running the Government Data Network in the United Kingdom. The initiative was first proposed in October 1990.

18 In the United States, the VSAT market experienced strong growth in 1984 with the AT&T divestiture and new opportunities to establish bypass networks. But by 1986 several of these – Federal Express Zapmail, a high-speed document transmission and electronic mail service based on a network of thousands of VSAT terminals – had failed. By 1989, there were 16,000 two-way VSAT terminals installed in the United States. The trend was away from systems based on one-way terminals to interactive networks, and most suppliers had learned to provide 'cradle-to-grave' service.

19 Following a Green Paper in November 1990, the Council adopted a Resolution on 19 December 1991 which set goals for satellite policy in Europe. These included liberalization and harmonization of the market, mutual recognition of licences, separation of regulatory and operational functions and improved access to space segments of international satellite organizations.

20 For example, in the United Kingdom in 1991 there were 1.1 million cellular radio subscribers, 55 percent of whom were self-employed; 20 percent were businesses with more than 100 employees, and 25 percent were businesses with fewer than 100 employees (Dataquest Europe 1991).

21 The British cable telephony market had reached some 72,681 installed lines customers by October 1992 and was estimated to be growing at 500 percent per year.

22 The steps in British telecommunication liberalization are found in Gist 1990. Oftel is a non-ministerial government department and is self-financed from licence fees. The Director General of Telecommunications must be consulted in the issuing of licences by the Government.

23 Certification procedures were moved to the British Approvals Board for Telecommunications (BABT) within the British Standards Institute.

24 See Licence granted to Cable & Wireless plc under Section 15(1) of the British Telecommunications Act 1981 (Mercury), 21 October 1982, superseded by Mercury Licence, 3/63/1/146.

25 British Telecommunications Act 1981: General Licence under Section 15(1) for Telecommunications Systems used in providing Value Added Network Services, 21 October 1982; draft licences published 13 June 1985 and July 1986.

26 In 1984, a Branch Systems Licence was introduced to stimulate the development of private networks, but only in so far as they did not involve the use of leased lines to construct networks that would handle voice traffic. In May 1987, the General Licence for Value Added and Data Services abandoned distinctions between basic and value-added services, liberalizing all services except voice telephone and telex. Under its licence, British Telecom is obliged to offer its network for use by other providers of value-added voice services (premium services for pre-recorded information). The difference between the price charged to the user and the local call rate is split equally between British Telecom and the information provider. However, competing service providers were required to use British Telecom's or Mercury's facilities and they were prevented from reselling PTO capacity. Fair trading provisions, similar to those in the British Telecom licence, were imposed on companies with an annual communication turnover greater than £1 billion; that is, they must demonstrate that no cross-subsidies exist. Large service providers were required to offer access to their networks and to implement Open Systems interfaces to prevent the supply of equipment becoming tied to the provision of services. By July 1989, this system of licensing had become unnecessary, and the relevant conditions for competing service suppliers were incorporated into a new Branch Systems General Licence which set out provisions for use of leased capacity. Simple resale of leased line capacity was allowed for domestic services and included the resale of both data and voice circuits.

27 It was not until February 1990 that British Telecom and Mercury connected their public data networks by high-capacity fibre optic trunks. The two different implementations of X.25 were to be handled by the X.75 protocol, which is also used to interface between various national X.25 networks.

28 An 'accounting rate' is a term referring to communication traffic between geographical areas controlled by different PTOs. The rate is established bilaterally and used to establish international accounts for revenue-sharing. The rate is expressed as a charge per traffic unit.

29 British Telecom's profit margins on international calls are about 60 percent, but it made operating profits of £2.3 billion on turnover of £9.2 billion from domestic services in the year to end March 1990 (Dixon 1990a).

30 'Equal Access' is a term first introduced in the United States by the Department of Justice for the divestiture of AT&T. Effective from 1984, it required that the RBOCs offer an equal quality of connection, at equal rates, to all common carriers. The term has subsequently been used to refer to a variety of different qualities and types of technical configuration for access to public telecommunication networks.

31 In order to strengthen competition, others suggested that British Telecom should be restructured into separate parts. For example, Ellison suggested that local transmission, local switching, long-distance transmission, long-distance switching and international services could be unbundled. This proposal called for British Telecom to publish separate accounts and to provide details as to how costs of overheads were allocated among them. Each business would retail its services through separate subsidiaries (Ellison 1990). Garnham also called for the structural separation of British Telecom's operations in order to differentiate between the role

of providing services to domestic customers and the role of British Telecom as a global company. He suggested that a new structure might consist of one entity to operate the local network; a second to provide a direct service to customers (e.g. terminal equipment, minutes of calling time, repairs, etc.); and a third to run the trunk network and British Telecom's other interests in manufacturing and international ventures. 'The aim would be to create a mosaic of local companies, and effectively restructure parts of BT' (Garnham 1990: 19).

32 British Telecom and other fixed operators were restricted from providing mobile or Telepoint services under their main licences, subject to review in the future. British Telecom, other national PTOs and Kingston Communications were restricted from conveying entertainment services in their own right until ten years after the publication of the Government White Paper. The possibility of reconsideration in seven years was mentioned; see Department of Trade and Industry 1991.

33 The Centre National d'Études des Télécommunications undertook much of the R&D leading to the introduction of packet switching, videotex, ISDN and other advanced services in France.

34 Broadcasting and other media services are provided under the responsibility of the Télédiffusion de France, and regulatory authority, from 1986, rested with the Commission Nationale de la Communication et des Libertés (CNCL). This organization authorizes radio spectrum use, cable networks and competing broadcasters though it can only make recommendations as to the extent of competition which should be permitted. Although it was to have taken up telecommunication regulatory responsibilities, these powers were returned to the Ministry of Posts and Telecommunication as a result of a decision in 1989. A Commission Supérieure de l'Audiovisuel now regulates the broadcast sector.

35 The CNCL did take responsibility for awarding concessions for private network operators during this period (see note 34). These networks excluded the private networks of a public authority and included terrestrial telecommunication installations exclusively in buildings or on territory owned by the operator which could be freely installed and operated; terrestrial telecommunication installations using buildings and land owned by a third party would need authorization; installations requiring the use of radio frequencies requiring authorization and the interconnection of private networks using leased lines with authorization. Interconnection of a private network with a public network was not, in principle, to be allowed.

36 The Décret du 24 septembre 1987 relatif aux liaisons spécialisées et aux réseaux télématiques ouverts à des tiers (Decree of 24 September 1987 with respect to specialised services and telematics networks open to third parties).

37 Two categories of services were introduced: specialized services – e.g. services which automate a function for the user, such as a network between an electronic funds transfer terminal and a mainframe computer or services for closed user groups – and non-specialized services which needed authorization. Services were those which needed more than 3.5 Mbit/s (or specialized services needing more than 5 Mbit/s). Specifications for international standards would be met to encourage open systems interconnection.

38 LOI no. 90-1170 du 29 décembre 1990 sur la Réglementation des Télécommunications (Law No. 90-1170, 29 December 1990 Defining the New Regulatory Framework of the French Telecommunications Industry). See also LOI no. 91-648 du 11 juillet 1991 and LOI no. 91-1323 of 30 décembre 1991.

39 The public status of France Télécom was regarded as making it difficult to respond to the needs of the marketplace. At least one observer of telecommunication in France has suggested that user groups have not been very influential, at least not in an organized way (Dang N'guyen 1988). From the perspective of one large user company, the combination of discretionary powers on the part of the minister, the fact that different accounting systems can be used to determine the costs of competitive services, and constraints on the transmission or transport component of competitive services which since 1987 could not exceed 15 percent, have not been problematic. In addition, the France Télécom approach to involving small and medium-sized firms in the development of videotex applications has helped to create an impression of an open market for the provision of informatics services.

40 Until 1984, the France Télécom subsidiaries were organized under France Cables et

Radio, a private company in which the state was the sole shareholder. In 1984, COGECOM (Compagnie Générale des Communications) was created as a holding company for fully owned France Télécom subsidiaries and other companies in which the France Télécom group holds shares up to 50 percent. Because COGECOM and the subsidiaries operate under private law, France Télécom has been able to engage in international activities that are prohibited for a public administration. Management flexibility was increased in areas related to personnel management, finance and marketing.

41 There is a different set of regulations for each type of service: telephone services between fixed points, including telephone booths in public areas and the telex service, are exclusive to France Télécom; the support services – e.g. Transpac or Numéris, which France Télécom can provide as it desires are open to 'controlled competition' and can be granted a ministerial authorization on a case-by-case basis, subject to specification sheets; radio-communication services – e.g. radio telephone – and telecommunication services supplied over the cable networks – e.g. remote monitoring – as a general rule require prior ministerial authorization; other 'information services' or 'value-added' services are unregulated and require declaration for prior authorization if lines are leased from France Télécom (France Télécom 1990: 7).

42 The general rule for establishing independent networks, designed for private or shared use, is freedom to establish. A licence is required when networks exceed 300 metres or they operate over 2.1 Mbit/s. The provisions for value-added services were established in March 1992.

43 The old PTT Code Article L-33 stated that 'no telecommunications infrastructure may be installed or used without the authorization of the Minister of Post and Telecommunications; this applies to sending and receiving any radioelectrical signals'.

44 A 'bearer Service' is defined as the commercial provision to the public of a simple data transmission, meaning a service including either the transmission, or the transmission and routing, of signals between telecommunication network terminals without these data being subject to any processing other than that required to ensure their transmission and the routing and processing associated with the control of these functions, see LOI no. 90-1170, 29 décembre 1990.

45 The latter figures include those using the French Télétel service, whereas there is no counterpart to this phenomenon in the United Kingdom.

46 Interview with France Télécom, 13 March 1990.

47 Agreements were signed bilaterally with the PTOs in Belgium, Spain, the United Kingdom (British Telecom), the Netherlands and Canada.

48 The legal background for DBP Telekom, apart from the general stipulations in the Basic Law for the Federal Republic of Germany, consisted of the Postal Administration Act and the Telecommunications Installation Act. The former laid down the responsibilities of the Minister of Posts and Telecommunications, other ministers, mainly Finance and Economics, and the Postal Administration Council which consisted of parliamentarians, employees and representatives of economic sectors. This group was able to insist on co-operation by the DBP on financial matters, operational decisions, tariffs and usage conditions, but it was dependent on information provided by the DBP.

49 In 1985, bilateral talks between the German and US governments resulted in concessions aimed at opening the market. Pressures to gain a larger share of world-wide equipment markets stimulated Siemens to take a more positive view of the potential liberalization of the domestic market in Germany.

50 This does not apply to the operation of telecommunication installations for the purpose of transmitting speech for third parties, which is the exclusive right of the Federal state. Initial exceptions to the network monopoly included the liberalization of two-way satellite communication at speeds above 15 kbit/s for data transmission in 1989. As early as 1988, discussions were under way to authorize PanAmSat, the American-owned satellite service provider to offer services between the United States and Germany.

51 All other services, including data transmission, have been liberalized and can be offered using leased circuits. Private service providers are required only to register their services and foreign ownership is not restricted. Resale of capacity is permitted as long as it does not involve a voice service.

52 In the case of mandatory services, if the competitive position of DBP Telekom as compared to firms offering the same or a comparable service were to be impaired as a result of its obligation to provide services, and compensation between monopoly and competitive services is impossible because of the lack of profits in the monopoly segment, competing enterprises may have obligations imposed relating to conditions of supply, e.g. quality or coverage, or prices.

53 These pricing measures, aimed originally to prevent 'cream-skimming', were withdrawn for low-speed, 64 kbit/s, and 2 Mbit/s lines, by July 1989.

54 By March 1991 it was clear that some twenty-three 'private' telephone networks had been operating in eastern Germany, and these were to be given permission to continue to operate for one or two years before being subsumed into the public network. These networks were estimated to support about 300,000 lines as compared with the 1.5 million for the whole of eastern Germany. This discovery was made when the western German chemical industry connected itself to the east Germany chemical industry network (Goodhart 1991b).

55 This expenditure aims to complete the installation of 7.2 million telephones, 68,000 public telephones, 360,000 fax connections, 50,000 packet switched data network connections, and 300,000 cellular mobile connections (Hiergeist 1991).

56 Ordinances apply to Swedish Telecom in certain areas, and these have been used to determine the quarterly subscription fee for residential telephone service, metered charges and duration for local calls. All other tariffs are set by Swedish Telecom (Martin-Lof 1989).

57 This service competes with Swedish Telecom's Nordic Mobile Telephone (NMT) system. By mid-1989, the Swedish Telecom component of NMT had 300,000 users, while COMVIK had only 20,000.

58 Although STN had a remit to monitor technical and standardization work, another specialized body, Swedish Standards Institution (Standardiseringskommissionen i Sverige) had direct responsibility for information technology and telecommunication.

59 Competition among the Nordic PTOs created difficulties for the venture. Sweden's approach to the international market was more aggressive than its Nordic partners. For example, the President of Swedish Telecom International noted at the time STI was created that 'we intend to use tariffs as a fancy weapon' (Gilhooly 1990).

60 Interview with Swedish Telecom, 28 February 1990.

61 Among the services considered for standardization of the required interfaces in Europe are freephone, credit card calling, virtual private network, organization, administration and management, and Universal Personal Telecommunications (European Telecommunications Standards Institute 1990).

# 5

# The Intelligent Network in the United Kingdom

> BT plans to become 'the most successful telecoms player in the world by the end of the 1990s'.
>
> (Vallance, British Telecom Chairman, November 1989)

British Telecom acquired the second largest American value-added network services operation, the McDonnell Douglas Tymnet network systems division, in 1989.[1] In doing so, as the trade press put it, British Telecom 'broke the unspoken covenant among Europe's telecoms administrations that forbade them from entering one another's territory' (*Communications Week International* 1989). In Tymnet, British Telecom argued that it had acquired the following competencies: substantial expertise in managing and operating public and private networks in the United States; entry into a world-wide market for services such as code, speed and protocol conversion, and the integration of different operations in a common facility; the capacity to develop gateways to ensure interoperability between dissimilar computers to reduce the processing load on customer mainframes; and the ability to offer cost-effective alternatives to leased lines for transaction processing applications and for sending data to multiple locations.

British Telecom announced in 1990 that it would team up with MCI to offer a service called International Featurenet and to offer a link into MCI's virtual private network in the United States.[2] Later in the year, another announcement pointed to a new one-stop-shopping service with a single point of contact and the option of paying bills in a single currency with twenty-four hour management. This service was called Global Network Services (GNS).

In the spring of 1991, British Telecom announced Pathfinder, a business which was initially to be 48 percent owned by the company (using British Telecom's Tymnet for US services), with NTT of Japan and DBP Telekom each taking shares. Pathfinder soon became known as Syncordia, a wholly owned British Telecom new company based in Atlanta. This company was expected to provide customers with bespoke solutions for their voice, data and video requirements.[3]

British Telecom's Chairman described the company's acquisition strategy as being one of expanding the customer base and the sales force internationally, rather than of acquiring infrastructure in foreign markets.[4] The strategy was clearly geared towards the global market, and the company would need to ensure that the domestic infrastructure could support its multinational customers' needs.

## British Telecom's Strategy

British Telecom had retained its dominant position in the UK marketplace despite the liberalization of the telecommunication market. It was British Telecom's vision of a flexible, customer-driven public network, capable of supporting the requirements of all types of users, that was expected to guide the parameters for the evolution of new services. British Telecom argued that it would respond to the forces of competition by turning towards the customer who would come first in all the company's endeavours. This was the Idealist vision. However, the Strategic reality was a continuing tension between the supply-led vision of the network engineers and the need to create a network capable of withstanding the incursions of new entrants. This environment left little room for responsiveness to *all* customer requirements.

Towards the end of the 1980s, British Telecom engineers and technicians adopted a more customer-friendly vocabulary. Network design would be *customer-facing*. The R&D division was expected to sell its services to other divisions which were interacting directly with customers. The orientation towards the demand side of telecommunication network design was being driven by the company's aim, which was to become an international manager of telecommunication networks.[5] In 1989, international business accounted for only about 10 percent of British Telecom's activities. By the early 1990s, the company was hoping to increase this proportion to 30 percent.

British Telecom had started its programme of expanding the capacity of its domestic network in the late 1970s. The company began purchasing multimode optical fibre in the late 1970s and, by 1983, was using fibre in all new trunk circuits. By 1989, 65 percent of the trunk network was comprised of fibre links. However, the extension of the bandwidth capacity of the trunk network had done little to put greater flexibility or control of bandwidth into the hands of the company's customers. British Telecom controlled customer access to the network, and the expansion of capacity did not necessarily imply the rapid introduction of new services.

Some pre-intelligent services were on offer prior to British Telecom's formal consideration of the intelligent network architecture. For example, the Linkline freephone service was launched in 1985.[6] In 1989, the company implemented an overlay network to support intelligent services such as freephone, credit card calling and advanced information services. Plans for wide area Centrex and virtual private networks were also under consideration. These services did not meet the formal architecture specifications of the intelligent network which were being debated at the time in various fora in the United States. The trade press reported in 1988 that British Telecom would probably move in the direction of the intelligent network.[7] However, the company's R&D staff felt that the Multi-Vendor Interaction Group in the United States had not come close to establishing a usable network model that could fully support the intelligent network services. The AT&T solution was regarded as being insufficiently flexible to provide a general-purpose platform to support the full range of telematics services.

British Telecom's early implementation of a limited range of intelligent services was made possible by the penetration of CCSS7 links. By 1988, the United Kingdom had a relatively high penetration of these links.[8] It was therefore not surprising that the company took the lead in implementing an early proprietary version of the intelligent network. The long-term modernization programme called for full digitalization and availability of CCSS7 links at all levels of the network.

The British Telecom network consisted of three tiers (see Figure 5.1).[9] Advanced services would be introduced using AT&T's Derived Services Network and, later, the company's own intelligent network architecture. Plans for new service introduction called for the use of a centralized Service Control Point (SCP). This would be provided using a mixture of the AT&T 5ESS overlay switches, adjuncts to existing facilities in the network and a programme of continuing switch development. Separate computers controlling intelligent network functions would be linked to the AT&T switches by a proprietary AT&T network control protocol.[10]

One long-term plan, when the full digitalization of the network was completed, called for the location of Service Switching Points in the local switches rather than in the trunk or Derived Services Network layer. From British Telecom's perspective, the intelligent network would make possible the introduction of services such as call identification. But the network designs under consideration in the late 1980s were deemed to offer limited scope for introducing new services without significant further development of both the switches and the Service Control Point software (Eburne 1989: 1).

British Telecom's design specifications were shaped and limited by technical constraints in the development of software. By 1992 or 1993 the company hoped to solve these technical issues and to demonstrate the implementation of the intelligent network principle of service independence from the physical network. A programme was initiated to define and implement a new Service Control Point (Eburne 1989). The coexistence of the AT&T overlay network and a new intelligent network architecture was uncertain. The AT&T overlay network did not have a design which would enable the straightforward unbundling of service functionality for use by the company's competitors. British Telecom was working toward the specification of network interfaces with its main switch suppliers – GPT and Ericsson. British Telecom argued that it had chosen the AT&T solution because neither GPT's System X nor Ericsson's System Y (AXE) were capable of handling advanced applications such as the Linkline service in

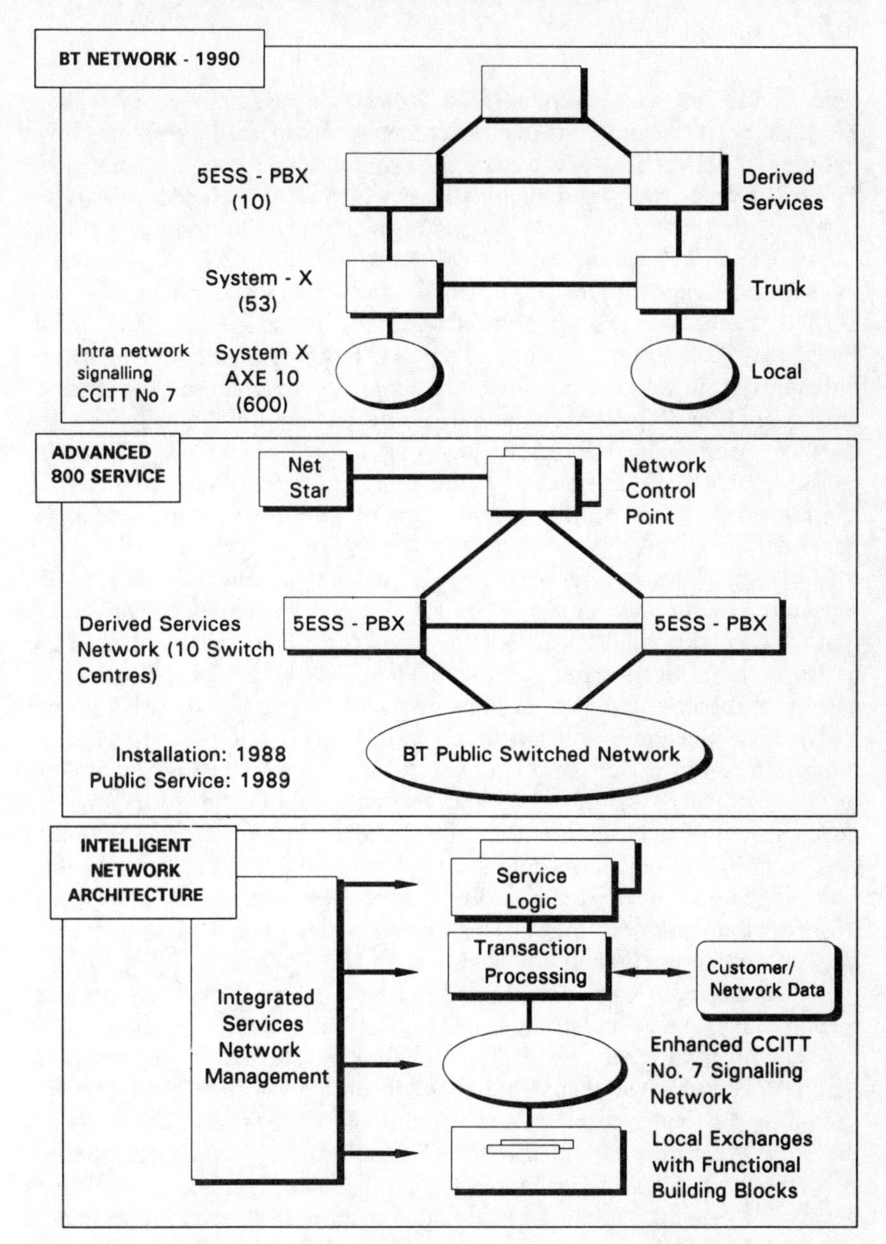


Figure 5.1 *The British Telecom intelligent network (Eburne, 1989)*

1989. In addition, virtual private networks, freephone and credit card calling had been running in the United States for a full decade before being introduced in Europe.

*The Engineering Challenge*

One engineering view within British Telecom's research laboratories at Martlesham at this time was that the company would need to retain control in order to deliver services to customers over a single integrated network. This was a vision of an intelligent network which would provide the *nervous system* of the new information age and it would depend upon operational software and the Information Network Architecture (INA).[11] This perspective on the transition to an intelligent network environment equated efficiency with network integration. Another important issue for British Telecom was the smooth evolutionary introduction of an intelligent network building on the existing network. Discussions with potential suppliers of equipment in both the telecommunication and computing industry focused on the appropriate architecture to support these objectives. There was considerable debate over whether the company's aims could be met via a centralized or decentralized distribution of intelligence throughout the network.

Martlesham Labs interviewees for this study believed that British Telecom would create an environment for service capability. This meant that the company would provide end-to-end functional management of the network. Some facets of network management would be moved closer to the customer, thereby increasing customer control.[12] Although British Telecom had a vision of a complex network fabric, it was very unclear as to what the interim steps towards it should be. Several interviewees believed that none of the company's strategic objectives could be met by pursuing the development of the intelligent network. There were many other potential design configurations that would meet the company's goals. The main issue was to decide what services would sell and what level of intelligence would be needed to support them. The other crucial issue was the level of interaction that the user should have with the network.

British Telecom was using laboratory-based service creation environments and experimental trials in 1989. The work was at an early stage. Demand for intelligent network functionality and services was described as being conceptual. The technical issues concerning how intelligent features should be physically distributed and the number of Service Control Points that would be used were still unresolved. Within British Telecom there was a concern about a gradualist approach to the evolution of an intelligent network. It was felt that the software modifications required to support new services would take years to develop. In the meantime British Telecom might be reduced to a *bit transporter*. As a consequence, the company could be bypassed by more sophisticated suppliers who had learned how to exploit advanced telematics networks. British Telecom felt that the answer was to move rapidly to exploit existing network investment to create an open, unbundled, distributed architecture, and to provide one-stop shopping services for the biggest customers.

In order to ensure this phased introduction of the intelligent network,

British Telecom entered discussions with several suppliers. The issue of the centralized versus decentralized location of intelligent network nodes became an ongoing debate (Debenham 1989). For some: 'Many of the services that could be delivered by means of an IN [intelligent network] solution are already available or will become available prior to the initial implementation of IN. In many cases alternative implementation strategies will exist for new services that are candidates for IN implementation' (Debenham 1989: 2). The company's ambivalence about open architectures and the creation of a multi-vendor environment for intelligent information services was in evidence in an attempt at collaboration with IBM. In 1989 British Telecom completed a joint study with IBM to evaluate the scope of the Application Programming Interface (API) in the Service Control Point of the intelligent network (British Telecom 1989). This interface was a specification which would enable a programmer to write service features that could be installed on any vendor's Service Control Point without detailed knowledge of the underlying hardware or software. This study proposed that British Telecom should manage the intelligent network centrally and that the Service Control Point should be supported by access to management functions and an open interface. For example: 'All management and administrative interfaces should conform to BT's OSI [Open Systems Interconnection] conformant Open Network Architecture (ONA). In addition, management of the SCP should comply with the BT–Open Network Architecture Management (ONA-M) requirements, so that management of the SCP is independent of the internal architecture or manufacturer' (British Telecom 1989: 15). The degree of customer's control over service creation would be restricted: 'Customers should have the ability to update a limited range of data without directly accessing the IMS (Integrated Network Management System) (e.g. directly via the telephony terminal). The IMS may hold a master copy of the relevant customer data, and provision must be made for this copy to be automatically updated on a timed or IMS initiated request basis' (British Telecom 1989: 18).

Despite the report's use of terms such as 'Open Network Architecture', British Telecom's assessment of the study determined that the IBM Application Programming Interface was vendor-specific and would need modification before it could become vendor-independent.

British Telecom was also involved in a European Commission project which aimed to develop a Functional Reference Model for a unified open services architecture. This was to be suitable for the development of public standards for an integrated broadband network environment (Kay et al. 1989). In this case, the long-term vision was to provide multiple service capability using network flexibility and intelligence. The network would be centred on an optical core with fixed delivery, access to a mobile periphery and provision of end-to-end functional service management. To support new services it would be necessary to introduce centralized network control. This architecture was addressed to the need to tailor customer services to rapidly changing needs.

*The Marketing Challenge*

One of British Telecom's major aims was to use the intelligent network to capture more of the market for the management of corporate networks. In interviews, the company's network designers said they wanted to create the equivalent of the 'friendly village operator' who was knowledgeable about all the attributes of a particular corporate customer. This meant that the company would need to retain control over network-generated information. By 1993, GPT's System X switches would be connected with a Service Switching Point using a proprietary interface. The architecture would allow the connection of more than one supplier's Service Control Point. A truly open intelligent network would require the public operator to have knowledge of customer attributes, to support access authorization and in all probability to relate personal data to a single number.

Intelligent network services had become closely linked to concepts of user mobility and the development of virtual private network segments of the market by early 1990. These were intended to counter the perception of a growing threat to British Telecom's revenue base from private networks and mobile communication operators. The company acknowledged that development of these services ran the potentially dangerous risk of opening the network to third-party operators. Whether the network would really be accessible to third parties was deemed to be a regulatory matter. However, it would also depend on whether the advanced intelligent network could be broken down into modular components. Thus far, all of the intelligent service implementations had been achieved using proprietary designs.

British Telecom's strategy stressed the ability to manage access to the intelligent network, and this approach did not exclude third-party use of network functionality. The company faced a recurring dilemma. If it spent to develop new systems, the way would be open for third-party suppliers to generate profits using the new functionality. If it did not invest heavily in new network resources, it would be unable to offer services equivalent to its major rivals in the international market.

Network development strategies were designed to overcome this dilemma. For example, British Telecom's Delta network plans were responsive to the requirements of large corporate customers, and at least one interviewee expected that they would take the 'wind out of the sails' of intelligent network developments along the lines proposed for public networks in the United States. The Delta overlay networks were intended to integrate private networks and the public network using Advanced Service Units which would allow customers to manage numerous sites. In this Delta model, intelligence was embedded within the switches. The model did not emulate the logical independence of the intelligent network from the physical network which was being championed in the trade literature and by the RBOCs in the United States. One view was that the equipment suppliers with the greatest interest in the intelligent network were those with the least functionality embedded in their switches. British Telecom would not wait for them to respond.

Instead, British Telecom's strategy would be to use the intelligent network as a gateway between numerous private overlay networks and public networks via an open interface. The company chose the Delta configuration because of the need to respond immediately to corporate demand for intelligence and flexibility over a seamless communication network. As a result, many proprietary signalling and management systems were being put in place linking multiple sites.[13]

Although network modernization was moving ahead in the late 1980s, the company's view of customer requirements seemed vague and hardly indicative of a demand-led response. One internally commissioned British Telecom study divided the potential market into four main segments: the *strong searchers*, who were usually businessmen who needed information handled in sophisticated ways over a mobile or fixed network; the *carers*, or people charged with the responsibility of looking after other people; *families and households*; and *everyone else* with much less need![14]

At least fifteen years, according to this study, would pass before multifunctional intelligent plug-in terminals would be located in residential premises. These would come much more quickly to the City of London. Although British Telecom was able to put multiplexing equipment on customers' premises allowing them to reconfigure their networks, there would still be an enormous need for greater network management capability. The company wanted to spread the intelligent functionality outward from a narrow range of business customers to small businesses and residential customers. But British Telecom interviewees were in some doubt about what services could be sold; how they would appear to the user; how the user would interact with the intelligent network; and how British Telecom would undertake research into user demand for new services.[15]

By the mid-1990s, British Telecom's representatives expected that the intelligent network would be targeted at two markets: personal mass markets and corporate customers. The primary functions would be call routing, access interfaces for mobile communication, removal of some of the functionality currently being installed in private networks and, finally, the creation of interfaces for the public intelligent network. 'In time the intelligent network would search out the existing functionality in the Delta Network so as to create seamless communication between the worlds of public and private networks.'[16]

The development of services like Centrex and virtual private networks could move in two different directions (Ralph 1988). Virtual private networks in the British Telecom/AT&T sense were directed to the need for managed connectivity between business sites without using physically dedicated channels. In contrast, Centrex would allow companies to use switching and call-processing functions in the public switches as if they were part of their internal private networks. This approach could be based on switches such as the Northern Telecom DMS 100. The result would be that British Telecom would diversify away from its dependence on AT&T's 5ESS switches.

British Telecom decided upon a strategy which was based upon moving quickly to proprietary standards in order to introduce advanced services for its largest customers using both public and private network configurations. The company believed that the process of reaching agreement on standards, whether with respect to the Bellcore version of the intelligent network, or a European standardized version, would be very slow. A move to an open network environment would require the company to have much greater knowledge of customer attributes in order to control access authorizations, and this would require the resolution of signalling and protocol issues. As a result, the seamless, open, telematics platform would have to be achieved via overlay networks targeting the business customer as a first priority.

British Telecom's impression was that neither AT&T nor Northern Telecom were particularly interested in the intelligent network because of the open interfaces. The company suggested that since these manufacturers have a commanding share of the market, the last thing they would do is to open the interfaces. In contrast, GPT and Ericsson were considered to be very anxious to develop open interfaces because this could give them an opportunity to increase their market share.

British Telecom also argued that intelligent network services would develop much more quickly in a monopoly environment. Their reasoning was that, in the more competitive environment of the 1990s, the company was concerned that its investment would be used by others to reap profits. Although open interfaces had yet to develop, the main impetus for the open network approach came from the RBOCs. Their aim was to gain independence from AT&T and Northern Telecom in order to push down the costs of switching equipment. British Telecom suggested that as the RBOCs entered the UK market through personal communication networks and cable networks, they would use their own intelligent network specifications. This would increase the eventual costs of service to the customer since interconnection with the British Telecom network would require the development of special gateways.

**Equipment Manufacturer Designs and Strategies**

In the United Kingdom, British Telecom turned to AT&T to supply its pre-intelligent network products, initially for freephone service. But the company continued to work with GPT and Ericsson to develop equipment for long-term network development. These designers had many interesting insights both as to their own goals for the intelligent network and the design criteria that would guide British Telecom and other British service suppliers in their development of more flexible networks.

*The GEC Plessey Telecommunications (GPT) View*

GPT had recognized that, to pursue its corporate strategy for growing equipment sales in an intelligent network environment, the main objective would be to develop equipment to support multiple service providers. To achieve open access to network intelligence would be an absolute necessity. Service providers would need access to databases to offer directory inquiry services, emergency numbers and personal numbering systems. GPT suggested that, realistically, the degree of access for different types of users would be a reflection of politics, rather than of technical potential and software development. The company expected that distinctions would need to be drawn between the public network operators and other service providers. They thought that confusion would arise wherever public network operators chose to provide intelligent services in competition with others. Because of these conflicting goals, the development trajectory for intelligent networks was very controversial in GPT in early 1991.[17]

The company envisaged a three-layer market structure. The first layer would be the telecommunication network operators, including the PTOs, the mobile and cable operators, who would install and operate the infrastructure and open the network to use by service providers. The second would be the service providers, who would both provide and use the underlying network resources. Network operation would need to be separated from service provision. This would ensure that public network operators would not gain unfair advantage when they also offered services. The third layer would be service retailing. Here services would be bundled by retailers, who would then sell tailored packages to the customer.

In GPT's view, the more entrenched, monopolistic PTOs would develop centralized intelligent networks or *classical* solutions. GPT argued that the design of intelligent network equipment would probably have to follow this route. But the company was more interested in the *adjunct* approach. This was regarded as being characteristic of the US market and a more open and liberalized telecommunication environment.

The classical solution – namely, the separation of intelligence from basic switching – had emerged in the United States and represented the route that AT&T had taken to provide its advanced freephone services. GPT regarded this solution as being suitable for a very narrow range of rudimentary applications such as the modification of call set-up procedures. This was limited to simple transaction services and was believed to be one area in which computer suppliers had a role to play.

But by early 1991, GPT saw signs of a new orientation to the intelligent network. This adjunct approach was premised on a close architectural link between the network intelligence and the switches. The classical model involved considerable latency or delay in signalling in the CCSS7 links operating at each node and prevented the emergence of a wide range of applications. The adjunct model called for a platform for rapid service delivery. In this area, GPT believed that, with the possible exception of Tandem Computers, computer manufacturers would have difficulty in providing the real-time software. Only Tandem and DEC had successfully supplied Service Control Points in the United Kingdom.

GPT's strategy was orientated towards the new decentralized intelligent network or adjunct model. The objective was to develop a model in which intelligence is more closely attached to the switches. In the new model – with intelligence as an adjunct to the switch – services could be delivered before CCSS7 was diffused throughout the network. Demand for services could be realized by embedding intelligence in switches distributed at different points in the network. This strategy was similar to that believed to be being adopted by Ericsson.

In this model there would be two layers: the first would be the service layer or Service Control Point based on a network-wide distribution of applications. The second would be the delivery, or service management system, supported by the interaction between the switch and a database. GPT expected that the interfaces for this configuration were very likely to be proprietary.

In the United Kingdom, echoing British Telecom's desire to maintain control of the intelligent network, GPT suggested that the intelligent network was a strategy for unbundling telecommunication applications software. GPT had collaborated closely with Tandem Computers to develop services for the initial British Telecom intelligent network pilots. The approach here was the classical one using a stand-alone Service Control Point. The orientation was not to the business market but rather to develop televoting and enhanced freephone services for the residential customer. The model was a semi-proprietary solution using a Tandem Service Management System connected to GPT's System X via an interface. The physical connection which requested instructions was proprietary.

GPT's representatives expressed interest in decentralized solutions which could be used to offer services such as teleconferencing. Tandem Computers would supply the database applications and GPT would develop the call control applications. Although GPT would collaborate in the development of operating systems, databases and user interfaces, it would produce the hardware itself rather than enter into an alliance with a computer hardware supplier.

GPT saw the main market opportunities for the intelligent network in three areas: adjunct services, service nodes (service logic in the switch), and private networks (for example, virtual private networks and Centrex). The company did not believe that virtual private networks and Centrex would be offered over the public intelligent network since large private networks were already using intelligent network technology. The company believed that this was simply a response by the PTOs 'to offer a comparable level of functionality'. In GPT's view, British Telecom would not attempt to meet large user needs through the integration of private and public facilities. Rather, it would meet the needs of private network users with an array of specialized networks and services which were separate from the public intelligent network. The only context in which British Telecom might be able to claim to be offering a seamless world of private and public networks was where the Northern Telecom Meridian PBX equipment was in use.

However, if steps were not taken by GPT to develop all alternative technologies – from public switches to private network solutions – businesses would use available bypass technologies to avoid using British Telecom's and other suppliers' local plant. GPT believed that British Telecom would bypass itself using high-capacity fibre optic systems, flexible primary rate multiplexers and network management systems such as the Delta network.

The US market was perceived to be much more liberal and conducive to the development of decentralized *adjunct* intelligent services. However, GPT recognized that there was considerable differentiation here as well. For example, US West was deploying the new adjunct architecture, whereas NYNEX and other eastern RBOCs were implementing services which were closer to the original or classical intelligent network concept. The result was that features could not be interworked across networks.

The fact that the price per line of public network switching technology had dropped from approximately US$400 to $150 per line in the United Kingdom and the United States was regarded as having turned switches into commodities. This was diverting attention away from intelligent networks. GPT also recognized that the costs required to enter foreign markets were substantial.

> We now recognise that it takes the same effort to develop, maintain, sell, support and manage a PABX in France, at least, as it does in England, and that is £1 million to adapt a product; £3 million per annum to keep it competitive and £25 million per year to obtain and keep a 15% market share. (GEC Plessey Telecommunications 1989)

The result of this trend had led GPT to launch a plan to break into the American market through its joint venture with Stromberg-Carlson in 1989. The Stromberg DCO switch had been designed for rural exchanges. By combining the power of the GPT System X with the design of the DCO, the aim was to create a marketable product for urban large switches – and to build in intelligent network functionality. The two key features under development were Centrex and virtual private networks.

By December 1990 British Telecom had endorsed the GPT System X approach to the development of intelligent network services. This move left AT&T without a market for its largest 5ESS switches in the United Kingdom. AT&T began promoting alternative technologies in the United Kingdom which could support more extensive use of private networks.

### *The Ericsson UK View*

Interviews with Ericsson UK representatives suggested that they regarded British Telecom's strategy towards the intelligent network as a move to forestall the development of more private networks built around PBX configurations.[18] Ericsson had experienced serious difficulties in meeting its obligations to supply the System Y (AXE) exchange to British Telecom. There were late deliveries, delays in developing new software, and problems in exchanging information which Ericsson considered proprietary with

British Telecom. The local System Y exchanges had to interwork with an array of analogue Crossbar and Strowger exchanges that were still in operation. The need to interwork with existing plant was perceived as a major cost factor. Another major problem was that equipment had to be designed both to interwork with proprietary signalling systems,[19] and to meet international CCITT standards.

Ericsson UK expected that costs would increase as a result of the existence of multiple suppliers to a PTO such as British Telecom. For example, the company estimated that the presence of a third switch supplier in the same market would lead to an increase in software maintenance costs which accounted for some 70 percent of all maintenance costs of digital switches.[20]

## Dominant Network Designs and Strategies

GPT and the other manufacturers of telecommunication switching equipment in the United Kingdom believed that computer and data-processing companies would assume that entry into voice communication product markets would be easy. They would soon find it was difficult. Eventually, telecommunication suppliers would need to cover the complete value chain, from products through systems to applications services, and even network operation. Features would no longer differentiate the offerings of the major manufacturers. The quality and range of products on offer would be essential. Factors such as account management, cost performance, user documentation, reliability and all aspects of *total quality* would be the main determinants of which firms would survive and grow in the 1990s.

In the United Kingdom, few at British Telecom or among the equipment manufacturers expressed concern about the need for standards prior to the implementation of intelligent network services. In fact, British Telecom had proceeded towards early implementation using proprietary systems embedded in the proprietary equipment in its network.

With respect to service implementation, British Telecom espoused an open network philosophy which would put greater control into the hands of its largest customers and enable the company to procure equipment from competing manufacturers. However, the company had stopped short of implementing an intelligent network design which would give clear signals to its competitors. British Telecom seemed preoccupied by the need to provide virtual private network and/or Centrex offerings to its largest customers. The company's understanding of wider consumer demand for intelligent services was very ambiguous.

The predominant strategy was one of centralized development of the intelligent components of the telecommunication infrastructure. This design was intended to support the requirements of large business users by a variety of both private and public network configurations. The evolutionary approach espoused by British Telecom involved an overlay network approach.

The evidence from the switch manufacturers suggests too that, despite a more competitive environment, British Telecom had established specifications for the intelligent network design with little user input from customers or equipment manufacturers. The switch manufacturers appeared to have been forced to follow, despite their perception that higher costs are associated with the development of proprietary systems adapted to the British market. For example, GPT recognized the need to develop systems for export markets, but it planned a different design for the UK market. Similarly, Ericsson UK found that it confronted problems in redeveloping its AXE switch for the local exchange market in the United Kingdom. Both manufacturers took the position that multiple supply leads to higher development costs. Direct collaboration with the computing industry, except in a relatively few areas, had not been extensive when it came to the implementation of the new intelligent systems which lie at the core of British Telecom's network.

## Notes

1 Tymnet was established in 1969 by Tymshare to supply remote computer services, and by 1974 the company provided data communications transmission for thirty major companies in the United States. In 1984, Tymnet became a subsidiary of McDonnell Douglas. Its Tymnet packet switched network had local access ports in over 700 American cities and approximately seventy foreign countries. The public network was a virtual circuit switched network with more than 2,850 intelligent nodes interconnected by terrestrial leased lines, microwave and satellite channels at speeds ranging from 4.8 kbit/s to 56 kbit/s. The public network was driven by a proprietary software programme – Supervisor – and the terminal interfaces supported both IBM SNA and X.25 protocols. British Telecom paid US$355 million for the data communication and value-added services businesses. The largest operator was Telenet Communications Corp, the originator of commercial X.25 packet switched networks for public use in the United States and now owned by Sprint.

2 A virtual private network offers services using virtual circuits in which there are no end-to-end connections for the duration of a call. Messages are transmitted as packets of information which have routing and addressing information. These pass through the network and share transmission paths with other messages.

3 The company was to provide three core services: Managed Links, Managed Private Network Services (MPNS) and Network Operations Management Services (NOMS) (BT 1991). Managed Links–Syncordia takes sole responsibility for providing, maintaining and managing point-to-point links; MPNS will provide a fully managed, dedicated network that can support all or part of a corporate communications system with guaranteed levels of quality of service, flexible bandwidth and simple network reconfiguration; NOMS will allow access to a complete outsourcing service for private networks, including terminal equipment, circuits and staff. By using CONCERT, BT's network management system, customers have an end-to-end view of the elements of the network. The CONCERT architecture is based on Open Systems Interconnection standards.

4 British Telecom did, however, purchase a 25 percent interest in Belize Telecommunications in March 1988.

5 In 1990, BT spent a total of £288 million on R&D, a large proportion of which was orientated to network development in support of this objective. BT's R&D spending was 2 percent of its annual income in 1990.

6 By 1989, this service had some 3,000 domestic subscribers and approximately forty customers for the international service.

7 The Bellcore IN+1 version of the architecture.

8 By 1989, 6.2 percent of BT switching equipment had CCSS7 capability as compared to 6.3 percent in the United States and the very high level of penetration of 47.6 percent in France; see Killette 1991.

9 Customers would be served by 600 GPT System X and Ericsson AXE10 local switches and some 3,000 co-located and remote concentrators; fifty-three GPT System X trunk switches; and an overlay Derived Services Network consisting of ten AT&T 5ESS switches to handle the advanced intelligent services such as freephone.

10 In order to secure the contract for the Derived Services Network, AT&T had incurred the costs of adapting its technology which had been developed originally for its own Software Defined Network in the United States.

11 Interview, British Telecom Martlesham, 6 September 1989.

12 Synopsis of view expressed by British Telecom representatives, visit to Martlesham, 6 September 1989.

13 British Telecom's proprietary Digital Private Network Signalling System (DPNSS) standard for PBX-to-PBX communication over private links would have to be made to work with the CCITT Integrated Services Digital Network standard for interworking between PBXs (Q.931). DPNSS is used in the Delta network system and it is not close to the Q.931 standard.

14 Interview, British Telecom, London, July 1990.

15 Interview, British Telecom Martlesham, 6 September 1989.

16 Interview, British Telecom, London, July 1990.

17 Interview, GEC Plessey Telecommunications, 1 February 1991.

18 Interview, Ericsson UK, 23 March 1989.

19 For example, Digital Access Signalling System (DASS) 2 and Digital Private Network Signalling System (DPNSS). DASS 2 was the British Telecom proprietary standard used in mid-1980s trials of pilot Integrated Services Digital Network applications using the primary rate interface.

20 Including parts and training costs.

# 6

# The Intelligent Network in France

> France Télécom has demonstrated its international character with partnership agreements, foreign investments and participation in international organizations. In a global context of growing competition for telecommunications services and privatization of major networks, France Télécom has a two-pronged strategy: to confirm its position as a prime player on the international scene and to promote the harmonization of standards between European operators. . . . With its new public operator status . . . France Télécom must more than ever be ready to listen to its customers.
>
> (France Télécom 1990: 9, 41)

The recent history of telephony in France can be characterized as a symbiotic relationship between the national government and the monopolistic telecommunication operator. The term '*dirigiste*' has been used to describe this highly interventionist and technology-led approach to the expansion of the French telecommunication network and the introduction of advanced services (Morgan 1989). At the beginning of the 1970s the French telecommunication network was one of the least developed networks in western Europe. In contrast, by 1992, this network was one of the most sophisticated in Europe. Over 97 percent of households were connected in 1989, the penetration rate had reached forty-five main telephones per 100 population and the penetration rate for advanced signalling capability conforming to the CCSS7 international CCITT standard was high.[1] France Télécom was reorganized as a public trading organization in 1991, and was using the intelligent network to underpin the modernization of the public network and to maintain a leading role in shaping the French telecommunication environment.[2]

## France Télécom's Strategy

The best known of France Télécom's early advanced service initiatives was the videotex service, Télétel, which had built up a base of 4.4 million Minitel terminals by 1989. This service was provided using the Transpac network. Although the promotion of videotex had been estimated as having generated losses of the order of some FF 4.1 billion in some years, the company argued that calculations made by the Treasury did not include revenues generated by traffic carried by the Transpac network, the carrier of Télétel.[3]

In 1986 a pre-Integrated Services Digital Network (ISDN) service, Transcom, was introduced, followed in November 1988 by the launch of the world's first commercial operation of ISDN. In 1992, full national coverage was established by the ISDN network known as Numéris. At that time the public network claimed one of the highest digitalization rates in Europe. The digitalization task had been nearly completed by the early 1990s including the local switching component of the network. This achievement had cost FF 1 billion, of which half was said to have been incurred in the provision of interfaces at each of the switches already embedded in the network.

France Télécom's strategy during the 1980s had been to build demand for its advanced services through partnerships with suitable companies. By 1988 in the ISDN area alone there were twenty agreements with users, providers and installers of advanced applications including surveillance systems, electronic funds transfer and so on. Agreements also have included those with Bull to develop hardware and software services, and with Electronic Data Systems to bring ISDN to users through PBX-based networks.

France Télécom was offering a variety of advanced intelligent services as early as 1985.[4] These services were aimed at private as well as public network users. For example, leased line services such as Transmic and the early introduction of Digital Automatic Cross-connect Systems (DACS) technology were intended to give large users flexibility in planning and controlling their corporate networks.[5]

This strategy had been described as an attempt by France Télécom to lock out competition from private managed data network operators (Gilhooly 1989b). The company's longer-term aim was to integrate ISDN and Transmic services to provide virtual private networks for its large customers. Consequently, the company's plans for the intelligent network embraced all facets of the infrastructure, including the public switched telephone network, packet switched data networks and leased line connections. The strategy deliberately targeted the need to create hybrid networks for the largest customers. Eventually the company intended to use the intelligent network to incorporate not only Télétel, the videotex system, and the Transpac network but also the public switched telephone network. At this point, France Télécom would succeed in providing fully integrated services.

France Télécom also expected to see a continuing demand for public and private networks. As one spokesperson observed:

> It would be easy to have a network with only switched connections, but to use a hybrid solution is much more cost effective. With ATM [Asynchronous Transfer Mode] technology there is actually no difference between switched and non-switched connections, but it would not be good to have only switched connections. For large traffic between two points, dedicated links are a better solution. To mix different kinds of solutions is the best strategy for the future.[6]

### *The Engineering Challenge*

France Télécom had defined a relatively clear strategy by the end of 1988 for the development of a nation-wide intelligent network beginning with the launch of intelligent network services at the end of 1990. The company claimed that its implementation of Alcatel E10 switches would represent the 'first commercial use of a Bellcore-type system outside the United States. . . . France Télécom is the first to address the subject and is moving ahead aggressively to offer greater services in this area' (*Communications International* 1989: 8). The most important services from the company's perspective were enhancements to the 'Numéro Vert' freephone services, wide area Centrex, automatic call distribution, alternative billing and virtual private networks.

The company's approach to the development of a suitable intelligent network had been under consideration by CNET, its research and development laboratory, since 1984, and was to be closely linked to the existing pre-intelligent services (Battarel et al. 1987; Bregant and Kung 1990; Collet et al. 1984; Kung 1989). Figure 6.1 summarizes the way in which France Télécom's engineering strategy has evolved. In this model, the Service Switching Point is always located in a network node and performs operations under the control of the Service Control Point.[7] There are several segments of the Service Control Point or intelligent node machine. The intelligent node is subdivided into several parts as shown in Figure 6.1. The PCS-R is the real-time computer or Service Control Point, and the PCS-G is the management machine or Service Management System.

France Télécom's staff would be able to access the Service Management System directly. They would also access the front-end machine called the PCS-C or Commercial Service Control Point. In addition, the PCS-D would be used for service creation (Kung et al. 1990). It would be only the PCS-C that would be accessible to customers, and then only under certain control and security conditions.[8]

The plan was to enable the customer interface to be integrated with the videotex network so that customers would be able to update, and possibly read, simple statistics. In the long term, more sophisticated interfaces would be implemented to allow greater customer control (Kung et al. 1990). The interface between the Service Switching Point and the Service Control Point would be vendor- and hardware-specific; that is, the Alcatel E12 switch.

France Télécom suggested that the main reason for not strictly following the Bellcore intelligent network proposals was the need to develop interface standards that would ensure greater control over the network. France Télécom planned to retain a monopoly over the infrastructure element of the public network. Thus, full open access could not be provided to the Service Management System component, or directly, to intelligent network databases. As the company's representatives put it, if customers were allowed completely open access to the network, France Télécom would need to protect the network against unauthorized access. While it was technically possible for private service providers to control a Service Control Point, this depended entirely on the evolution of regulation.

The intelligent network was regarded as part of the 'bearer', 'basic' or transport network where very little value added is created. As long as the

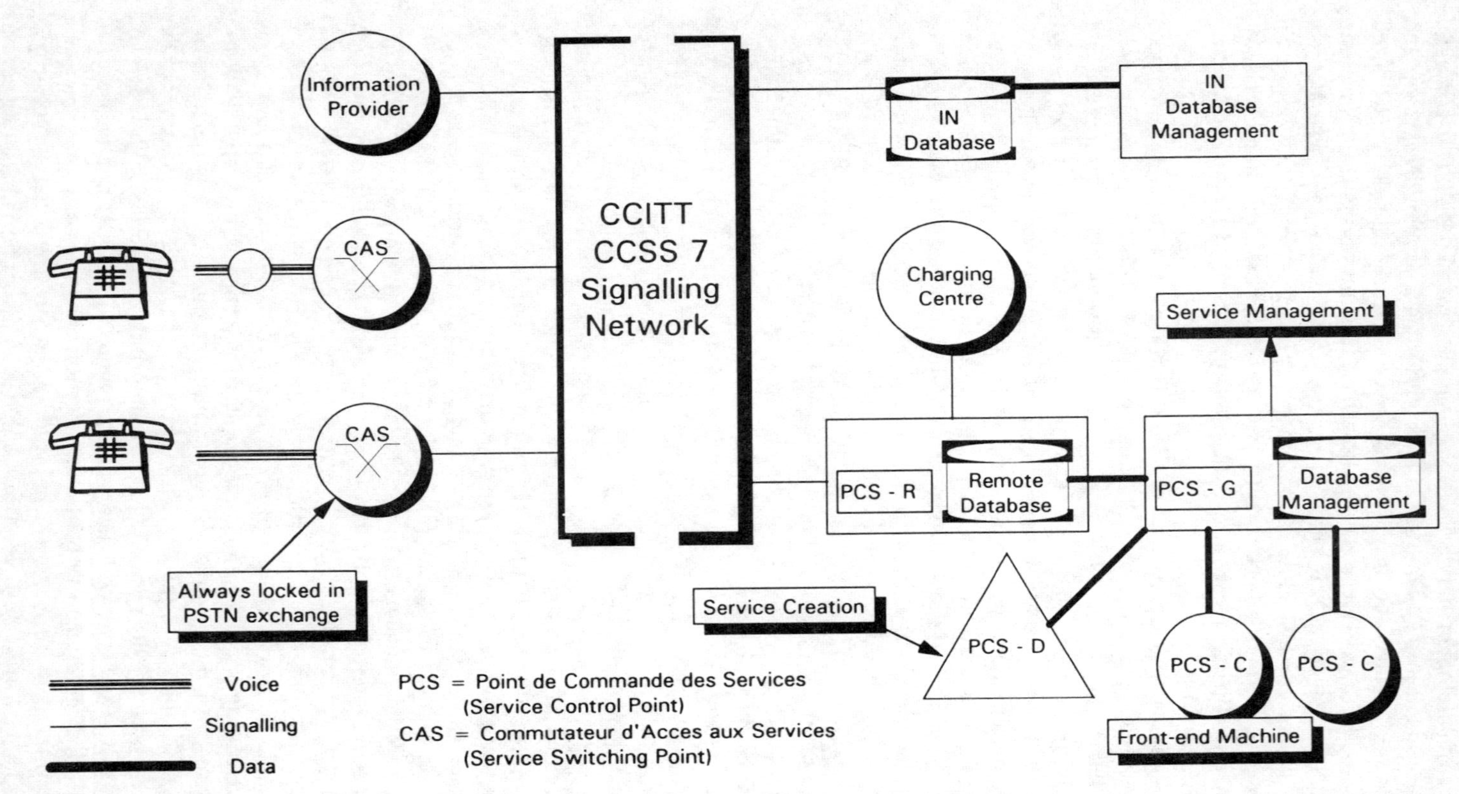


Figure 6.1 *France Télécom's intelligent network (International Switching Symposium, 1990; European Telecommunication Consultancy Organization, 1990)*

regulatory environment continued to permit monopoly supply of the public infrastructure, France Télécom would maintain a monopoly in intelligent network service supply. Company representatives suggested that the main reason was not the company's desire to protect the infrastructure from competitors, but rather to protect the customer from security failures and down time in the network. For one representative, all the issues surrounding the intelligent network were concerned with the level of control available to network operators and users.[9]

The challenge for network designers was to find an effective way of controlling the gateway between public and private networks (Kung et al. 1990). CNET was studying the possibility of building a barrier into the intelligent network Service Control Point to stop external entities from entering the public network and causing disruptions. The Service Control Point had to be under public control so that the overall integrity of the network could be maintained.

The France Télécom Service Switching Point was to be delivered by mid-1990 by Alcatel. The company planned to work with two different suppliers for the SCP-R, the real-time computing capability, and one other supplier for the commercial Service Control Point which would be accessible to customers. The Service Control Point for the enhanced Smart Card service was being developed by DEC and Sligos, a software company.

As of 1990 the France Télécom network was offering pre-intelligent services using Service Switching Points and Service Control Points located in some of the transit exchanges. These included Numéro Vert (freephone) and Carte Pastel (credit card authorization). By 1991, CNET expected that the Services Réseau, intelligent network services, would be introduced.

In the early phase of service introduction enhancements were to be made to existing services; for example, Smart Cards and virtual private network services. The additional benefits provided by the intelligent network were expected to include customized announcements, universal numbering, improved access to the network, freephone in an ISDN environment, call queuing and call distribution. The intelligent network modifications would also increase capacity and provide the basis for a variety of additional enhanced features.

Intelligent network implementation was also to be used to enhance the virtual private network, Colisée. This network was already supporting private numbering plans, volume discount tariffs and producing network statistics. Intelligent enhancements would include ISDN, access to international switches and user management systems.

Although France Télécom had been collaborating with computer and software companies, this appeared to have little to do with the intelligent network architecture itself. Company officials emphasized that these collaborations had been under way since the development of ISDN supplementary services such as sub-addressing and call-line identification which were undertaken with DEC, Nixdorf, Hewlett-Packard, IBM and Groupe Bull.

France Télécom was committed to building up partnerships with intelligent network service suppliers in order to stimulate the development of the French software industry. The company saw this as having been a successful strategy in the case of videotex and it was being pursued again. The goal was to achieve partnerships with respect to intelligent network services but only in so far as the services complemented those offered by France Télécom.

### *The Marketing Challenge*

Marketing perspectives within France Télécom differed from those of the visionaries who were designing and specifying the network. Representatives of the Marketing Department felt that the earliest introduction of intelligent services would be 1992 or 1993, and these would almost certainly not be based on the Bellcore architecture. Marketing representatives also believed that a key issue was the extent to which security would be jeopardized by allowing customers to gain access to the network.

They argued that the new equipment to support intelligent network services was at least three years further away from operational implementation than the estimates offered by technical designers. There had been serious difficulties in implementing the software for pre-intelligent services. The process was proving extremely costly, despite the fact that the benefit of the new design was intended to be cost reductions in implementing new services.

Initially, supplementary services like call-forwarding were being provided to users without charge by France Télécom. The take-off of these services would depend on how long the pricing policy allowed services to be offered at zero or minimal cost to the customer. Some representatives within France Télécom suggested that there was no obvious demand for virtual private networks. The real requirement was for several types of control, and this could be met using a variety of technical alternatives. The intelligent network architecture was not essential to meeting the customer requirements with which most of the marketing personnel were confronted.

More important was the need to find new ways to generate revenues in markets that had not been developed by competitors. Perhaps most important of all was the need to serve large business customers. The top 2,000 business customers accounted for 30 percent of France Télécom revenues. By 1990 the shift in emphasis toward the business customer had been publicly declared despite the agitation that this had caused within the organization and criticism from the unions. 'Intelligent services would be obviously aimed at the 500 biggest firms which needed to be kept loyal. It was hoped that in the long term advanced services would be of benefit to small and medium sized firms.'[10]

The issue facing France Télécom was how investment in the intelligent network would bring competitive advantage in existing and new service markets aimed at the largest firms rather than at the mass subscriber base. Company officials recognized that pressures for the interworking of the different components of national and international networks were being created by large users. However, they suggested that public networks would continue to be incompatible to some degree because of problems in agreeing standards. France Télécom was moving faster than the standards-making organizations in Europe and internationally. The company had no intention of slowing the development of new services. Proprietary implementations of intelligent services were regarded as inevitable, and the cycle of incompatibility for new generation equipment would continue as far as France Télécom interviewees were concerned.

The substantial R&D costs associated with public network switching technologies would also continue to grow. France Télécom argued that these components of the network would never be completely opened to competitive bidding. All switches had different capabilities and features which evolved to suit particular national environments.

**Equipment Manufacturer Designs and Strategies – the Alcatel View**

In some quarters, computer companies arc thought to be eagerly awaiting the opportunity to enter telecommunication manufacturing via designs for the intelligent network architecture. However, a tender by France Télécom for the Service Control Point had resulted in only three bids from Alcatel, DEC together with Sligos, a French software house, and Bull in partnership with Sesa, another French software house.[11] In this section Alcatel's perspective on the intelligent network is highlighted.

The first details of Alcatel's intelligent network strategy were revealed in 1988. The company aimed to develop intelligent network products without entering alliances.[12] The main emphasis was on the development of its System 12 and the E10 switches. The objective was to maintain both switches but to develop common components which would enhance their capabilities. Intelligent network products would be designed to create overlay networks and they would be based on Alcatel's 83 computer processor.

Like other European manufacturers, Alcatel initially announced that it would develop its intelligent network architecture on the Bellcore model.[13] Company officials emphasized that their approach to the intelligent network was a proprietary design based on their own fault-tolerant X.83 processor. This approach would distinguish Alcatel products from those of their competitors. Alcatel's participation in the development of the intelligent network freephone service for France Télécom was described as a 'telecommunication' problem, whereas the Smart Card, for which DEC had won the contract, was regarded strictly as a 'data-processing' problem. Alcatel argued that this division of design expertise was indicative of the role that data-processing firms would play in the future. Whenever the basic telecommunication network was involved, the design task would belong to the telecommunication equipment manufacturers. Alcatel was not intending to collaborate with companies such as IBM because computer manufacturers were not considered to have the competence to run telecommunication networks. Alcatel regarded itself as an equipment supplier, not as an applications supplier.[14]

One Alcatel representative interviewed for this study regarded the intelligent network as a way of introducing new subscriber services, of optimizing network management and of configuring virtual private networks on the public network. The company's approach was designed to enable PTOs to offer a wide range of services to business and residential customers by the end of the 1990s. In most cases, in Alcatel's view, initial services such as detailed billing information and call-forwarding would be orientated towards the residential customer.

The Alcatel perspective on the intelligent network located the architecture concept within a set of wider initiatives directed towards the future of broadband networking. The public switching systems division of Alcatel Network Systems defined broadband networks as 'those networks that support services requiring bit rates well above 1 Mbit/s' (Depouilly 1990). The E10 and System 12 narrowband digital switches were to provide the foundation for the evolution to broadband networks. The need to cope with the introduction of service applications flexibly and cost-effectively was regarded as the driving factor behind broadband and intelligent network architectures.[15]

Alcatel's designs attempted to distinguish between value-added services and services that are basic to the network. The company distinguished between two types of services:

> Services that involve extra resources in the control network, such as call forwarding, 800 [freephone] services, and central routing. Although they need special translation, routing or charging facilities in addition to those for normal call setup, they do not require any special treatment of the transported data.
>
> Services that involve extra resources in the transport network; these additional resources may be located within the network (e.g. videoconferencing, coding law conventions), if they are provided by the public company, or connected to the network as end users if operated by a private company (e.g. TV distribution). Transmitted data, in addition to being switched in the normal way, has to be specially processed in accordance with the service, or generated at the subscriber's request. (Beau et al. 1990: 141)

Figure 6.2 shows the main elements of the Alcatel architecture. In principle, network resources might be operated by the public-network operator or private-service providers. It was here that the intelligent network architecture would play a central role. According to Alcatel, the network design and the standards adopted by the manufacturers would be the determining factors in whether private-service providers faced low, or extremely high, entry barriers as they sought access to future public networks.

Alcatel representatives also argued, however, that the business customer had no need for ISDN or greater intelligence in the public network. Private networks would continue to be organized around leased circuits and packet

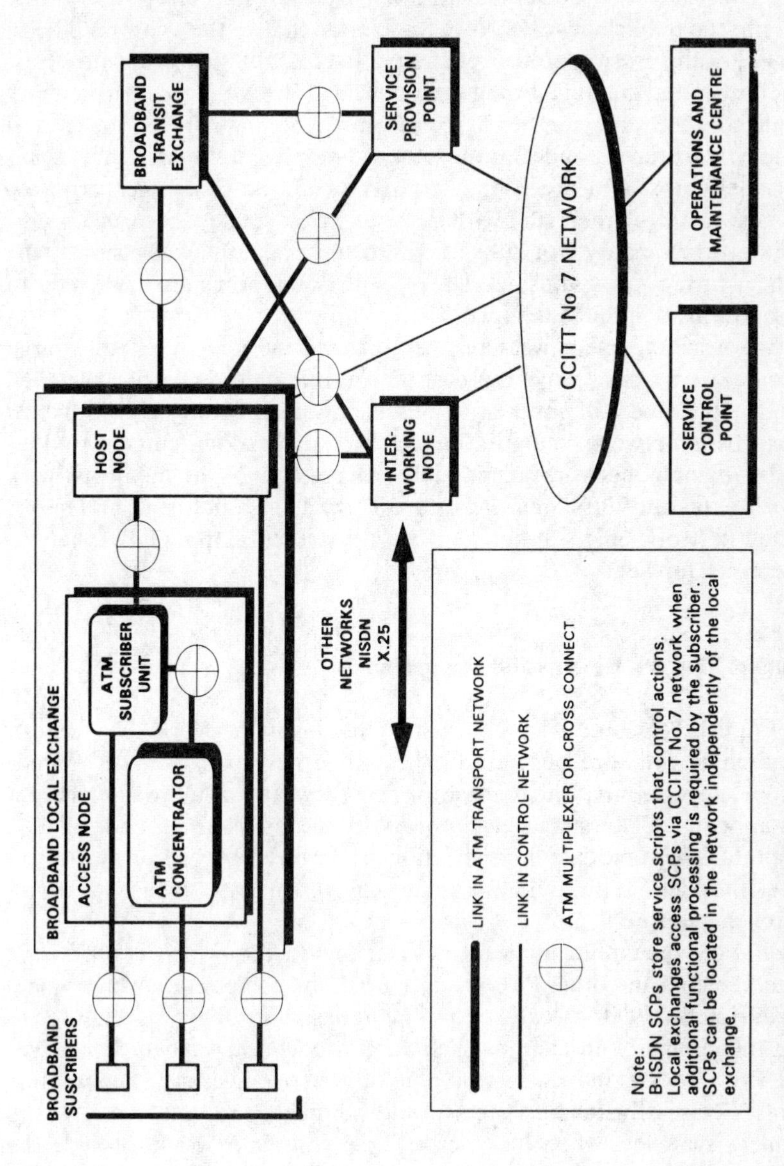


Figure 6.2 *Alcatel's broadband network architecture (Beau et al., 1990: 143)*

switching technologies.[16] The trend would be towards PBX-based core networks linked with the PTO's public networks to form hybrid solutions.

From Alcatel's perspective, there were two strategies with respect to the business communication market. The first would use public switched network services such as Centrex to link customers and bring their traffic back onto the public network. As in the United States, the aim would be to fight against the erosion of market share as a result of bypass technologies. Alcatel suggested that this strategy was typical of British Telecom's planning and reflected the large number of private corporate networks in the United Kingdom. In France, and Europe more generally, there were far fewer private networks, and the PTOs' strategy would be to enable corporate customers to link all their sites with PBXs and Centrex services. As a result, the same functionality would exist within the public network and at the periphery within various business sites. These two approaches would have to coexist within an intelligent network environment.

Network management was considered to be the key to whether large corporate users would move more of their traffic onto the public switched network. In France, in contrast to other European countries, the hybrid public/private network configuration was already predominant. Two other options – namely, networks completely independent of the public network using satellites and Very Small Aperture Terminal technologies, and public switched network-only solutions – were considered extremely unlikely to evolve much further.

**Dominant Network Designs and Strategies**

Like France Télécom, Alcatel saw the intelligent network as a way of strengthening the core business of the public network operator. France Télécom suggested that the market for new services was more open than it often appeared. The state's tradition of coherent planning had made it difficult to accommodate liberalization in France, but according to an Alcatel interviewee, progress was being achieved in a way 'which fitted with the French culture'.

Alcatel saw the intelligent network as an architecture which would enable France Télécom and other PTOs to face up to the competition which would come during the 1990s. Alcatel's equipment design strategy was regarded as giving the company an edge over competitors who had allied themselves with IBM.[17] Alcatel interfaces would be offered for a variety of computing systems. Service Switching Points would be located in both the E10 and System 12 switches, while the Alcatel X.83 computer would provide the Service Control Points and Service Management Systems functions.

This strategy had been aimed originally at capturing a share of the American market. However, by 1990 the company was ready to withdraw because of the high costs of modifying system designs. Like GPT in the United Kingdom, Alcatel believed that the intelligent network, and all facets of public network modernization, should be aimed at the mass consumer or residential market, rather than at the large business users. Finally, both France Télécom and Alcatel were strongly of the view that it would be the regulatory environment that would set the parameters for the type of functionality that would eventually be incorporated within public network intelligent switching systems.

## Notes

1 See National Telecommunications and Information Administration 1991: 177, 189, cited in Ergas 1992.

2 As an 'exploitant autonome de droit public' France Télécom remained within the public sphere but gained greater possibilities of pursuing commercial interests. For an overview of the earlier strategies of France Télécom within the Government's strategic industrial policies see Ergas 1992.

3 For a comprehensive recent review of the France Télécom Télétel services, see Jouet et al. 1991.

4 For example, the Numéro Vert service had 4,776 subscribers at the end of 1985 and 12,998 by the end of 1989. The number of calls charged grew from 16.4 million during 1985 to 60 million in 1989. A credit card service had been introduced in 1984.

5 DACS are used to interconnect low- and high-speed digital circuits and provide access to voice and data channels. In 1988, AT&T sold thirty DACS nodes to the company. Combined with software, this equipment could provide network management facilities in a similar fashion to the intelligent network configuration on the public network.

6 Interview with France Télécom, 13 March 1990.

7 The company's technical documentation refers to the 'Point de Commande des Services', the Service Control Point (SCP) in the Bellcore model, and the 'Commutateur d'Accès aux Services' or the Service Switching Point (SSP).

8 Interview with France Télécom, 13 March 1990.

9 Ibid.

10 Ibid.

11 IBM was not among the companies to whom tenders were sent. 'It's a problem of competition', said one France Télécom spokesperson.

12 Alcatel's spending on R&D reached ECU 1,344 million in 1989 or just over 10 percent of total sales.

13 The Bellcore version IN/2.

14 Interview with Alcatel, Paris, 14 March 1990.

15 Central to both is the concept of separating call control and connection control functions and support for interworking between different types of networks.

16 Interview with Alcatel, Paris, 14 March 1990.

17 The strategy is based upon the AT&T-derived UNIX operating system.

# 7

# The Intelligent Network in Germany

> The intelligent combination of network technologies and value-added services facilitates the use of synergies which are particularly cost-efficient, and at the same time makes it possible to find flexible customer-related solutions.
>
> (DBP Telekom 1991a)

In Germany there is an emphasis on developing a public telecommunication infrastructure that will serve the communication requirements of all users in the twenty-first century. The Integrated Services Digital Network (ISDN) and the intelligent network are centrepieces of this strategy. The main strategic question that appeared to be facing Deutsche Bundespost (DBP) was the balance between upgrading existing network resources and investment in new transmission and switching technologies.

Historically, the German PTO had the right to a monopoly in the operation of the public telecommunication network and to determining its technical characteristics. There had been few direct challenges to the monopoly status of the public network operator until the late 1980s. Few private networks had been established, and various pricing strategies had been introduced in an attempt to ensure that leased capacity would prove unattractive for most large users as an alternative to the public switched network. However, as early as 1978, there had been criticisms of the DBP's practices, which were restricting the emergence of advanced communication services. Led by companies such as Nixdorf and IBM, the first inroads of liberalization in the terminal equipment market began to be felt.

**Deutsche Bundespost Telekom's Strategy**

Faced with the continuing build-up of pressures to liberalize the telecommunication market, DBP Telekom's ambitions by the end of 1991 had become clear. The company wanted to be a 'gateway to the European networks', the importance of which would grow significantly in 1993 when the European Community became a single market (Ricke 1991: 6). The design of the public infrastructure would contribute to this goal by enabling the company to provide customers with the full-service options embodied in one-stop shopping, single-end billing and network management. These services were being offered as supplements using existing network capabilities.

Until the liberalization of the German telecommunication market, there had been little pressure on the company to develop intelligent network services. The DBP had introduced a freephone service, Service 130, in 1986,[1] and the company was offering a voice mailbox service as an additional telephone service feature.[2] Another feature associated with the intelligent network is call-forwarding, and this too was being offered without the formal introduction of the new intelligent network architecture.[3]

In 1990, Telekom's U2000 network nodes were being designed to permit switching and multiplexing functions to be performed by computers. The U2000 network initiative offered an alternative design to increase the intelligence in the public network. From the year 2000, about one-third of all local and long-distance lines were to be controlled and managed by these switching nodes, and eventually they would support virtual private networks and call number identification. As a result, the intelligent network architecture being debated in the United States was only one of several configurations which the company had under consideration at the end of the 1980s.

### *The Engineering Challenge*

Informal talks had been held between DBP and IBM, DEC, Bell Atlantic and Siemens in 1988 to formulate an intelligent network strategy, but few at the company had expected this to become a reality before the 1990s. Preparations for the intelligent network were intensified in 1990 and tenders were invited internationally for the network components (DBP Telekom 1991b: 78). The newly reorganized DBP Telekom had set the scene for a country-wide operational trial which was to begin at the end of 1992. This move was justified by the fact that, although most of the services under consideration already existed, the new configuration would improve the quality and range of service features. The technical advantages would include more rapid call set-up and connection and the separation of these functions from higher levels of intelligence in the databases.

DBP Telekom representatives emphasized that there already was an intelligent infrastructure in the long-distance public network and that databases using CCSS7 links were in operation. They suggested that the removal of regulatory restrictions on the use that the company could make of information contained in this database had stimulated its cautious movement towards the intelligent network concept. However, the intelligent network architecture was also considered to bring disadvantages. Traffic-flow management would become increasingly difficult for the public operator, with a concomitant increase in the risk of network failure. A centralized intelligent network model would inflate costs since this would require increased levels of redundancy in the public network. The DBP Telekom engineering perspective on the intelligent network was that it should simply be regarded as an evolutionary step in upgrading the public switched network.

DBP Telekom's philosophy in the early 1990s was to use the functionality of the public switched network to support non-voice services wherever this was possible. But representatives argued that 'it costs more to go into the higher end level' and there were no strong indications as to the direction in which user requirements would develop.[4] Once the investment was made, the market could easily go in another direction, and the company's preference was to avoid the substantial risks inherent in a strategy of rapid implementation of the formal intelligent network architecture.

Nevertheless, by 1991, DBP Telekom's investment strategy had begun to crystallize. The main focus was to be on intelligent networks with broadband structures, optical fibres to the home and digital mobile communication systems (DBP Telekom 1991b: 12). Emphasis was being given to the introduction of CCSS7 as a pre-condition for the development of intelligent services. Investment in this signalling system was expected to permit a clear separation between services and the underlying network. Service elements could then be offered transparently, and the need for private network investment on a nation-wide basis would be reduced.

DBP Telekom rejected the notion that the intelligent network architecture would permit a move away from the telecommunication switch manufacturers towards the computer vendors. The company argued that there would be a need to obtain limited assistance from computer manufacturers and software developers, but they would be involved only in the first installation of equipment. Thereafter, DBP Telekom would do software upgrades internally.

The company was aware that the computer companies were suggesting that public network operators would be able to modify software parameters. However, although Telekom maintained staff trained to make software modifications, there was no internal software development branch. The company had most of the required knowledge but not in sufficient quantities. Thus far, modifications had been done by the switch manufacturers. DBP Telekom interviewees suggested that if the computing companies seriously wanted to become involved in the telecommunication market they would have to offer continuing maintenance and support.

DBP Telekom planned to choose three manufacturers to equip three network nodes with Service Switching Points, Service Control Points and Service Management Systems. Tendering companies included Siemens, Alcatel SEL, Philips, Telenorma, Ericsson, Northern Telecom, AT&T and IBM. The winners were Siemens, Northern Telecom (with Cincinnati Bell Information Systems) and Alcatel SEL with DEC. They were to introduce four test services – caller identification, improved freephone service, televoting and automated customer billing – and field trials are continuing to 1995 at least. Virtual private network services were to be delivered towards the end of 1993 (Schoenbauer 1993). DBP Telekom had retained its monopoly only over the transmission network, and a link between virtual private networks and Centrex would mean that private services could be offered over the public network, challenging DBP Telekom's position in the market.

The company was taking the lead in specifying services and the interfaces for the Service Control Point and Service Switching Point functions. After this design phase, interviewees believed it would be possible to split responsibility for the development of the intelligent network between suppliers. To guarantee compatibility between different parts of the public network, the company would develop good working relationships with these suppliers. A balance would be struck between choosing among competing suppliers and engaging in collaboration to ensure compatibility.

DBP Telekom's network designers saw the value of the intelligent network architecture in the fact that it would reside behind the public switched network where the company could maintain control over access to the Service Control Point databases. Although private intelligent-service vendors might emerge, they would supply the applications markets. They would not compete directly with DBP Telekom. Competition might develop in database supply and applications. But deeper inroads into the core activities of the public network would depend on the access conditions. The biggest barrier was considered to be the absence of agreed interface standards. If these emerged there would be real competition for the supply of the Service Control Point and the Service Management System. The intelligent network standards that had been developed in the United States were not perceived to be sufficiently international. Interviewees suggested that these would not go unmodified in Germany, or indeed, in the rest of Europe. DBP Telekom would retain a strong position in value-added services markets because, even if competing services were to be permitted from a regulatory point of view to establish their own private Service Control Points, they would face substantial entry barriers due to the absence of standardized interfaces with the public network.

Telekom's intelligent network was being implemented in two phases. The first involved the installation of centralized network nodes to support initial intelligent services.[5] In 1991, three pilot projects were under way: caller identification; freephone service; and televoting and automated customer billing. The company expected the largest proportion of traffic carried on the intelligent network to be generated by large corporations. The second phase was planned for completion by the year 2005. By then, 180 trunk exchanges would be able to access intelligent network features. These trunk exchanges would use high-capacity signalling connected to database nodes. The costs of new service introduction were expected to be reduced because there would no longer be a need to modify all switches in the network to support new services.[6]

### *The Marketing Challenge*

DBP Telekom interviewees suggested that there were no plans to determine whether investment in the intelligent network could be justified in cost terms. The market was described as a 'sleeping beauty who only needed to be woken up'. In the early 1990s, there was much uncertainty among marketing personnel as to the development trajectory for service applications. Marketing personnel appeared to regard the intelligent network as a relatively straightforward enhancement of existing public network switching functions, not as a new phase in the evolution of networks. Though intelligent network resources would clearly be capable of supporting new data and image services, DBP Telekom's income was derived mainly from basic telephone services. Value-added services were described as 'butterflies', non-essential services that DBP Telekom might or might not choose to offer.[7]

The main application of the intelligent network would be in support of freephone services and signal re-routing, and the architecture was not essential to this. During the transition to the new architecture most of the development effort would focus on voice-related services, where there was more predictable demand.

DBP Telekom was very cautious about the potential demand for non-voice services which might be supported by the intelligent network. Nevertheless, interviewees acknowledged that the company would probably attempt to provide a wide spectrum of non-voice, value-added network services. It would also bid to plan, manage and operate networks for large corporate users. Most intelligent network traffic would be generated by large business users, especially by routing traffic between offices and plants.

## Equipment Manufacturer Designs and Strategies

### *The Siemens View*

The Siemens vision of the intelligent network in the late 1980s was of a service-independent network architecture allowing the simple introduction of new services. It was the clearest illustration of an engineering-led vision. One spokesperson forecast that by the year 2000 everyone would have 'the ISDN in the home'. Broadband would be provided as an overlay network since it was not needed by everyone. Public and private narrow and broadband networks would be much less integrated than the engineering visions would suggest. The Siemens approach to the intelligent network was part of its vision of 'more intelligence, bandwidth, flexibility, mobility, economy and convenience for the benefit of users, service providers and network operators' (Siemens 1990). The model was a close fit with the company's long-term development strategy, which envisaged a universal network for all services. The company's vision was driven mainly by large and medium-sized companies, public authority and research sector requirements for special data and image applications – for example, coupling, local area networks, metropolitan area networks, document transfer, video conferencing, and so on. From the network design perspective, Siemens looked forward to the realization of private switched and dedicated networks within the public network. The company believed that this approach was initially gaining significance for companies operating from several locations, but also for external transactions with partner organizations – for example, between manufacturers and their wholesalers (Siemens 1990: 13).

Company representatives suggested that three main services would be embedded in the public network:

1 interactive high-speed data communication providing connectionless services, i.e. local area networks operating at 10 to 135 Mbit/s; connection-orientated services running at 2 to 135 Mbit/s, and document transfer and retrieval services running at anywhere from 2 Mbit/s to as fast as 600 Mbit/s services;
2 interactive video communication for high-quality graphics, videoconferencing at 64 kbit/s to 135 Mbit/s, video telephony at 64 kbit/s to 135 Mbit/s, and broadband videotex and video retrieval;
3 distributed video communication for television production and conventional and high-definition television distribution.

The Siemens model for the intelligent network would allow existing switches in the public network to be upgraded to Service Switching Points or to Service Control Points via the addition of software. The Service Management System would be implemented on the Siemens 7500 and MX computers.

As early as 1988, Siemens had observed that a key problem in any public network strategy would be the relationship between the ISDN and the intelligent network. The problem for the network designers was whether these two models would create synergies for service implementation or compete with each other (Maher et al. 1988). The fascination with the intelligent network was attributed to a 'vision of a flexible network fabric which enables an economic, timely, and ubiquitous provisioning of services' (Maher et al. 1988: 316). However, although the benefits, possible architectures and evolutionary strategies had emerged by 1988, little if anything was known about introduction strategies or the ways in which network operators would ensure that different pieces of the technical puzzle would fit together.

In the Siemens strategy, the ISDN would provide user-friendly access to the public network. The intelligent network would provide a flexible set of packages of functionality.[8] Both would depend on the pervasive implementation of CCSS7 to provide appropriate signalling. Siemens regarded its EWSD public switch as an integrated package capable of providing the optimal technical solution. The components could be deployed throughout the network in whatever configuration the public network operator desired. The result would provide the ingredients for a transformation towards a *Universal Network*.

Siemens envisaged that competing service vendors would use the Universal Network. This would require open non-discriminatory access. For example:

> concepts of Open Network Architecture and Open Network Provision are aiming at an open non-discriminatory access (comparably efficient interconnection) of

> third party enhanced service providers to the basic network. More precisely, telecommunication networks must be able to offer basic network resources and functions to Value Added Service providers on an *unbundled*, i.e. *individually tariffed* basis. (Frantzen et al. 1989: 153; emphasis added)

Siemens' network designers observed that by 1989 some 118 Basic Service Elements, the public network bundles of software functionality, had been compiled from service provider requests in Germany (Frantzen et al. 1989: 153). The intelligent network would also be used to improve the provision of basic services by introducing a more flexible network control architecture.

There were several alternative ways in which the intelligent network could be introduced, taking into account the availability of CCSS7 and ISDN access. Services could evolve from called party services such as freephone, to calling and called party services which depended on the directory number from which the call originates; that is, identification, and communication with the calling party (such as alternative billing, enhanced emergency services, virtual private networks and wide-area Centrex).

A third category, and the most interesting from the point of view of competitors, would be interactive services which would require dialogue capability with the user. This category could include interactive freephone services, voice messaging, call completion and private transaction networks, home banking and pay-per-view television.

However, this category presented the greatest difficulty. A definition of network access capability would need to differentiate between various types of network access (and costs) on the basis of the degree of service interaction available to the user. As a result, service interaction or control issues embodied in the standardization of interfaces and services would be 'perhaps the most critical requirement' of the intelligent network (Maher 1989: 67).[9] Figure 7.1 shows the points at which critical standardized interfaces would be needed (indicated by heavy black lines).

These standards were not fully in place, although International Telecommunication Union CCITT specifications were considered to provide initial definitions. Standards for service definitions – for example, processing requirements for billing and traffic measurement and for existing operations, such as administration and maintenance of systems – would also be needed. These would be crucial to the compatibility of public networks and these were furthest from definition.

From a competitor's perspective, another element of the intelligent network would be equally important. To allow for programmability of services via a standard language from an external source – that is, a public network operator's competitor – and to provide flexibility without sacrificing security, it would be necessary to provide an interpreter code:

> the success of IN [intelligent network] depends on the ability of operating company personnel to create new services – without direct knowledge of the implementation within any specific vendor's switch . . . a challenge will be to define a service logic language that is relatively efficient in all vendors' switches

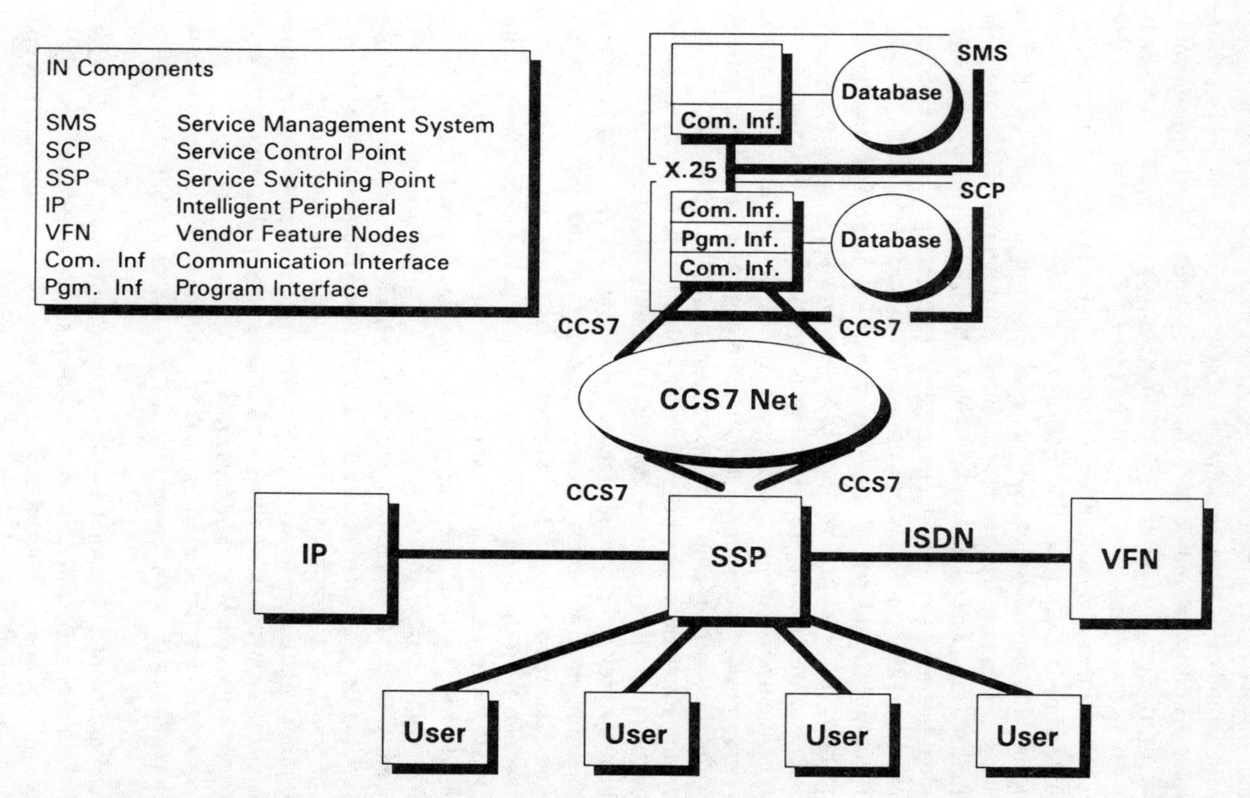


Figure 7.1 *Intelligent network technical overview (Ambrosch et al., 1989)*

without any unfair disadvantage to any particular switch architecture. (Pierce et al. 1988: 27)

This, too, was far from being fully developed.

Siemens gave central importance to the public network. The intelligent network would be a critical ingredient in the trend away from private networks towards the public network. Some dedicated private networks would survive, but they would become increasingly rare with the development of Siemens' public Universal Network. One official interviewee described the 'catastrophe' that would result if all networks did not become integrated into broadband public networks.

Turning the engineering vision into a reality had presented difficult problems for Siemens. The result of a joint initiative between Siemens, IBM and Bell Atlantic was published in 1989 (Ambrosch et al. 1989). The study had demonstrated critical blockages to collaborating with a computer manufacturer. Siemens completed the study convinced that IBM's weakness was its failure to understand that real-time processing is vital in telecommunication. Siemens argued that IBM was unable to understand that intelligent network nodes are critical for network operation. If they do not function, the whole network can be blocked.[10] However, Siemens had little prior experience of marketing or database management, and it was in these areas that the company gained the most from this collaboration.

Siemens emerged with the view that only the Service Management System would fall within the domain of computing companies. There would be a grey zone around the Service Control Point, but computer company implementations would not be strong on reliability or real-time functions. Siemens saw its own competence as being especially strong in the lower levels of the Service Switching Point. The company admitted that its EWSD switches were not completely decentralized or accessible to those who might wish to collaborate in the development of intelligent network components.

By early 1990, seven public switches in Germany supported intelligent network features. Siemens' strategy was to develop the basic EWSD switch and to seek collaboration for smaller intelligent network components. According to one company manager, 'if you are to be a number one player in the world telecoms switching market it is necessary to distinguish between basic products like EWSD and niche markets'.[11]

Siemens argued that if a switch manufacturer needs to collaborate in the development of the core technologies, 'then the battle has already been lost'.[12]

Siemens acknowledged that a multi-vendor environment was really only a vision because of the standardization problems. As another manager put it, 'there is always an own interpretation of standards . . . someone has to pull the network together, and the effort is proportional to the number of vendors'.[13] Thus Siemens' overall strategy placed standards issues at the core as a critical factor in the development of advanced equipment to support new public and private network services.

**The Alcatel SEL (Standard Elektrik Lorenz) View**

In contrast to Siemens, Alcatel SEL saw the biggest barrier to the development of intelligent network services as being the attitude of the DBP Telekom. DBP Telekom was not perceived to have the marketing skills necessary to create demand for the new services. Traditionally, SEL and DBP Telekom had 'paid little attention to user needs'.[14] One SEL manager noted that intelligence already resided in the public network. In his view, the intelligent network was simply another catchword to follow upon ISDN and should not imply that there had been a radical change in the capacity of the equipment manufacturers or the PTOs to respond to customer needs. 'In the past we were technology driven, and in the future we are market driven. This is at least what everyone says. But of course it is always a mixture and the target is to make business.'[15]

From SEL's perspective the intelligent network Service Control Point would be used to upgrade the freephone Service 130 and to offer virtual private network services. Since standards were not ready, the company would proceed with proprietary interim protocols. SEL observed that the introduction of new billing services would allow private-service providers to emerge. But if DBP Telekom were to retain its monopoly on advanced voice applications, then the intelligent network would be designed to support these and opportunities for competition would be reduced.

SEL compared the intelligent network evolution in 1990 with broadband technologies in the early 1980s. Just as there was no visible demand for broadband applications for the bulk of subscribers then, there was no demand for intelligent network services in the early 1990s. SEL even suggested that the enthusiasm for intelligent networks would fade away in much the same way as it had for broadband. Network evolution would be a function of the strategic objectives of the PTOs and the equipment manufacturers as it had always been. If there was a response to demand, it would be to large-user demand – 'Users do not know what they want and SEL does not know either.'[16]

SEL acknowledged that the company had argued publicly that a shift from hardware to software in telecommunication networks would mean that telecommunication companies would have to collaborate with computing companies. Companies like IBM had approached the PTOs to handle both the telecommunication and data-processing sides of the business. SEL believed that DBP Telekom had involved itself, for example, with IBM in order to push the telecommunication equipment suppliers to modernize. But the aim was simply to instil an innovative spirit in the traditional manufacturers, not to engage in longer-term collaboration. The telecommunication equipment manufacturers were striking back with their own proposals, and the computing companies had shown that they could not handle basic telecommunication network management functions.

## Dominant Network Designs and Strategies

The intelligent network was regarded by DBP Telekom as a way of increasing the volume of traffic over the public network. Tariff policy was widely regarded as being critical to the take-up of new services. Some interviewees suggested that there would be a need for a cost methodology that would attribute costs to different services in order to determine their viability.

Siemens' vision of a Universal Network looked forward to the further blurring of the technical boundaries between public and private services and saw this as being inevitable. For example:

> The old boundaries between public and private networks are becoming more and more blurred: components of the public network can be used for private networks, components of private networks can be used for public networks. The provision of public networks with varied and flexible uses can technically and economically weaken the position of private special-purpose networks outside the public network. (Siemens 1990: 22)

The intelligent network architecture would offer a solution to the public network operator to ensure that functions would work together. Standards would be needed to guarantee open, non-discriminatory access to the network by independent service providers, but these were not in place. Competition in the supply of advanced services would materialize only with the advent of 'a standardized provider-network interface with special signalling functions, but primarily the splitting of the network features into individually callable and chargeable elementary components' (Siemens 1990: 30).

The concept of access was central to Siemens' view of the intelligent network. Interviewees suggested that without standardized interfaces for access, the new architecture would be no more open than earlier network configurations.

Alcatel SEL was of the opinion that the development of the intelligent network depended on DBP Telekom's future ability to market its services. The company was not convinced that demand for most applications was in place or that manufacturers and the PTOs were responsive to demand except, perhaps, by the largest users.

## Notes

1 By the end of 1990 some 3,600 Service 130 numbers had been assigned, an increase of 50 percent over the previous year. In July 1991, the company reduced the prices for this service in order to attract business customers. By July 1991, there were approximately 5,800 customers.

2 The number of voice mailboxes had reached 1,080 by the end of 1990.

3 GEDAN call forwarding; by the end of 1990 some 22,500 call-forwarding facilities were in operation.

4 Interview with DBP Telekom, Bonn, 19 February 1990.

5 These nodes are the Service Control Point, Service Switching Point and Service Management System; see *Communications Week International* 1991a.

6 For example, before the development of the intelligent network Service Control Architecture, it was not possible to separate the set and release function of switching from higher levels of intelligence, i.e. the supplementary services contained in databases. Every time a new service was introduced, the software in approximately 6,200 local switches in Germany had to be modified.

7 Interview with DBP Telekom, Bonn, 19 February 1990.

8 This included Broadband ISDN, Asynchronous Transfer Mode, higher-layer functions for value-added services, and the Telecommunications Management Network (Frantzen et al. 1989).

9 CCSS7 provides the link between the Service Switching Points and the Service Control Points. In Siemens' view, the implementation of CCSS7 according to common standards was one key to a multi-vendor environment. The transaction-based communication protocol between the Service Switching Point and the Service Control Point would also need to be standardized to ensure that calls could be routed to the Service Control Point.

10 For example, if the response time of the Service Control Point is slow, the network must recognize that it cannot handle calls and instruct the Service Switching Point to re-route traffic if the whole network is not to be blocked. Several Siemens interviewees believed that this was the main reason for the crashes in the AT&T network in early 1990; interview with Siemens, 22 March 1990.

11 Interview with Siemens, 22 March 1990.

12 Ibid.

13 Interview with Siemens, Munich, 21 February 1990.

14 Interview with Alcatel SEL, Stuttgart, 22 February 1990.

15 Ibid.

16 Ibid.

# 8

# The Intelligent Network in Sweden

> The overall goal of the telecommunications policy is to ensure that the general public, business, industry and public agencies throughout Sweden have satisfactory access to telecommunications services at the lowest possible cost to society.
>
> (Televerket 1990)

Like other public network operators in this study, Swedish Telecom has aspirations to become an international operator. This objective has been stimulated as a result of the considerable number of globally operating firms based in Sweden. It has also become a strategic necessity in the face of the gradual build up of competition to the company's *de facto* monopoly. Swedish Telecom is facing competition especially from Tele2, which is 60 percent owned by Industriforvalnings AB Kinnevik, an investment company that has moved into the information-services business, and 40 percent by Cable & Wireless. In 1993, Sweden was only the second European country, following Britain, to authorize international competition in the supply of voice services. Swedish Telecom has advocated its own restructuring to allow it the flexibility to compete alongside other internationally operating companies.

Thoughout the late 1980s and early 1990s, Swedish Telecom had been responding to the growth of private networks and competition in its service markets. The company had relinquished its *de facto* monopoly control over the provision PBXs and modems in 1987. The managing director of Swedish Telecom had observed that 'it is still our ambition to be a leading supplier of private branch exchanges in Sweden, but it is essential to trim back the organization for sales, installation and administration and to cut jobs' (Webb 1987). PBXs had been used to support the development of corporate networks. However, these networks were believed to be generating substantially less revenue for the company than would have been the case if the same traffic had been built up using the public switched network.

### Swedish Telecom's (Televerket) Strategy

The Swedish telecommunication network had reached a relatively high level of digitalization by the end of 1988. Over 50 percent of the trunk network and 33 percent of local exchange network were composed of digital facilities.

The previous year, in advance of any clear demand for a new public network service, Swedish Telecom had introduced a circuit-switched 64 kbit/s network offering public and private data transmission. The introduction of this network, together with the extensive availability of CCSS7, gave the company a different perception of the urgency of introducing the new intelligent network architecture as compared to its counterparts in the United Kingdom, France and Germany. With the advanced signalling system already in place, the company was able to begin to experiment with the introduction of advanced service applications.

Swedish Telecom interviewees suggested that the intelligent network really began to be introduced with the installation of Stored Programme Control (SPC) switches in 1980. These switches had provided limited local intelligence in the public network, but when this was combined with the early start towards implementation of CCSS7 in 1983, the company had been able to experiment and to introduce very advanced services.[1]

Freephone services had been introduced in 1987. Calls were translated in the local AXE switch, and the company had plans to centralize the number translation functions. As advanced services grew in volume, the plan was to install a database in each geographical region. A Centrex service for smaller PBX users had also been introduced in 1987 as well as wide-area Network Centrex for larger customers.[2] The Centrex service was built on top of the AXE switch and was partly a response to competition from private networks. An information-provider service which used an access code for a new scheme of call-charging was introduced in 1989. However, ISDN would only be installed in the 1992–99 time frame with commercial availability from 1995.

### *The Engineering Challenge*

Swedish Telecom was working with Ericsson on the development of the intelligent network. Company officials believed that Ericsson had a 'bee in its bonnet' with regard to the fact, that when a new service was introduced it would need to be centralized at locations within the network. The Ericsson view was that only after demand for a service had been demonstrated should the service intelligence be decentralized throughout the network.

From an engineering perspective, decentralization of network intelligence would be necessary because of the need to cope with huge volumes of transactions and to build reliability into the network. From Swedish Telecom's perspective, any service architecture which would induce premature obsolescence of existing switches was to be avoided and, for this reason, Ericsson's modular intelligent network design was regarded as a good strategy. But company officials observed that 'the reality is that modularity might not be as easy to achieve as Ericsson believe'.[3] An intelligent network infrastructure that might attract larger customers onto the Swedish Telecom public network was only beginning to be implemented in 1990. It was generally agreed by company representatives that it would be some time before these developments would challenge the private networking capabilities of the largest customers which had been built up over the past decades.

Although Ericsson, together with Swedish Telecom, had co-operated in the development of the AXE public network switch and the MD110 PBX, other manufacturers were nudging at the edges of the Swedish market. Siemens, Philips, Alcatel, IBM and Nippon Electric had all expressed interest in establishing a stake as suppliers in the Swedish market. Swedish Telecom was distributing Northern Telecom's SL-1 exchanges under licence and had done so since the 1970s. Though only approximately 20 percent of the equipment used in the company's network was imported in 1989, Swedish Telecom aimed to retain only the profitable parts of the telecommunication terminal equipment business.[4]

Swedish Telecom's intelligent network strategy was aimed at the public switched network and at the private PBX-based network market. The company advocated the introduction of capacity resale for leased lines to increase the attractiveness of private networks and it had not sought to protect its voice telephone market. Instead, the company argued that 'no distinction can be made between voice and data in a digital world' (Robotham and Walko 1989). The intelligent network strategy was embedded in the company's plan to introduce wide-area Centrex, a service that would expand local Centrex services and capitalize on the extensive PBX installed base that would be nearing the end of its service life during the early 1990s. As this occurred, private networks could be linked more effectively into the Centrex (switched) network, generating revenues and traffic for the public network operator, Swedish Telecom.

Swedish Telecom, through Teli, manufactured nearly all the PBXs required to meet its own requirements. It also supplied 70 percent of the AXE switches it needed for the domestic network. As a result, the company was not heavily dependent on Ericsson's equipment delivery plans. Nevertheless, Swedish Telecom was doing very little of the design work itself. The basic design work for the AXE was undertaken by Ellemtel, and the heavy development costs were being shared by Ericsson and Swedish Telecom.[5] The company had been more concerned with making modifications to Ericsson designs to suit domestic market needs. But Swedish Telecom's ability to carry out design modifications on the AXE switch had given it the opportunity to experiment with new services.[6]

Swedish Telecom embarked on its programme of digitalization in 1980, much earlier than Germany, France or the United Kingdom. The CCSS7 signalling system was installed beginning in 1980, available on a trial basis in 1983 and in full operation by 1985. Between 1985 and 1990, all the company's 135 AXE digital exchanges had been connected to the signalling system. In 1987, Swedish Telecom began implementing intelligent network applications by introducing Service Control Point switches into the voice network. By 1990, 40 percent of Swedish Telecom's subscribers to the public switched telephone network were connected to intelligent network nodes.

### *The Marketing Challenge*

Swedish Telecom managers were reluctant to accept the idea that the intelligent network would create the possibility to respond more directly to customer demand. As one manager observed, Swedish Telecom needs to be seen to be providing all types of communication. It should do this by maintaining a local presence and developing special contracts for the larger business users, not by implementing advanced network architectures that would simply increase costs.

The key problem was described as being to maximize network use and to build bridges between networks, rather than substantially to upgrade or integrate services within the public switched network. Flexibility from both the network operator's and the customer's perspectives could be improved in many ways, including the use of multiplexing techniques, ISDN, network management tools and so on. These did not necessarily need to be incorporated within the intelligent network framework. Several Swedish Telecom inteviewees suggested that 'it is not the case that everything has to be integrated, some things would be better specialized. Today it is still easier to solve user needs via special networks.'[7] They also noted that, despite progress towards digital intelligent networks, there was considerable analogue equipment embedded in the network. The company would want to protect this investment. It would not seek to replace it simply to introduce services for which there was very uncertain demand.

Figure 8.1 shows the telecommunication operating environment that would challenge Swedish Telecom and its equipment manufacturer in the 1990s. For Swedish Telecom the optimal intelligent network would minimize bypass and maximize the traffic generated by the installed base of PABX equipment.

## Equipment Manufacturer Designs and Strategies – The Ericsson View

> The objectives of the Intelligent Network can be met in a number of ways, and several different scenarios of how the network may develop can be envisioned. One view is that the service logic execution will always be centralized in general purpose computers: the switches are depleted of service intelligence and simply become transport switching devices. Ericsson considers this view to be simplistic, and fundamentally wrong.
>
> (Ericsson Telecom 1988)

This quotation highlights one of the differences between the Ericsson and Bellcore approaches to the intelligent network architecture. The intelligent network architecture was originally regarded as a perfect example of the synergies between computing and telecommunication (Soderberg 1991).

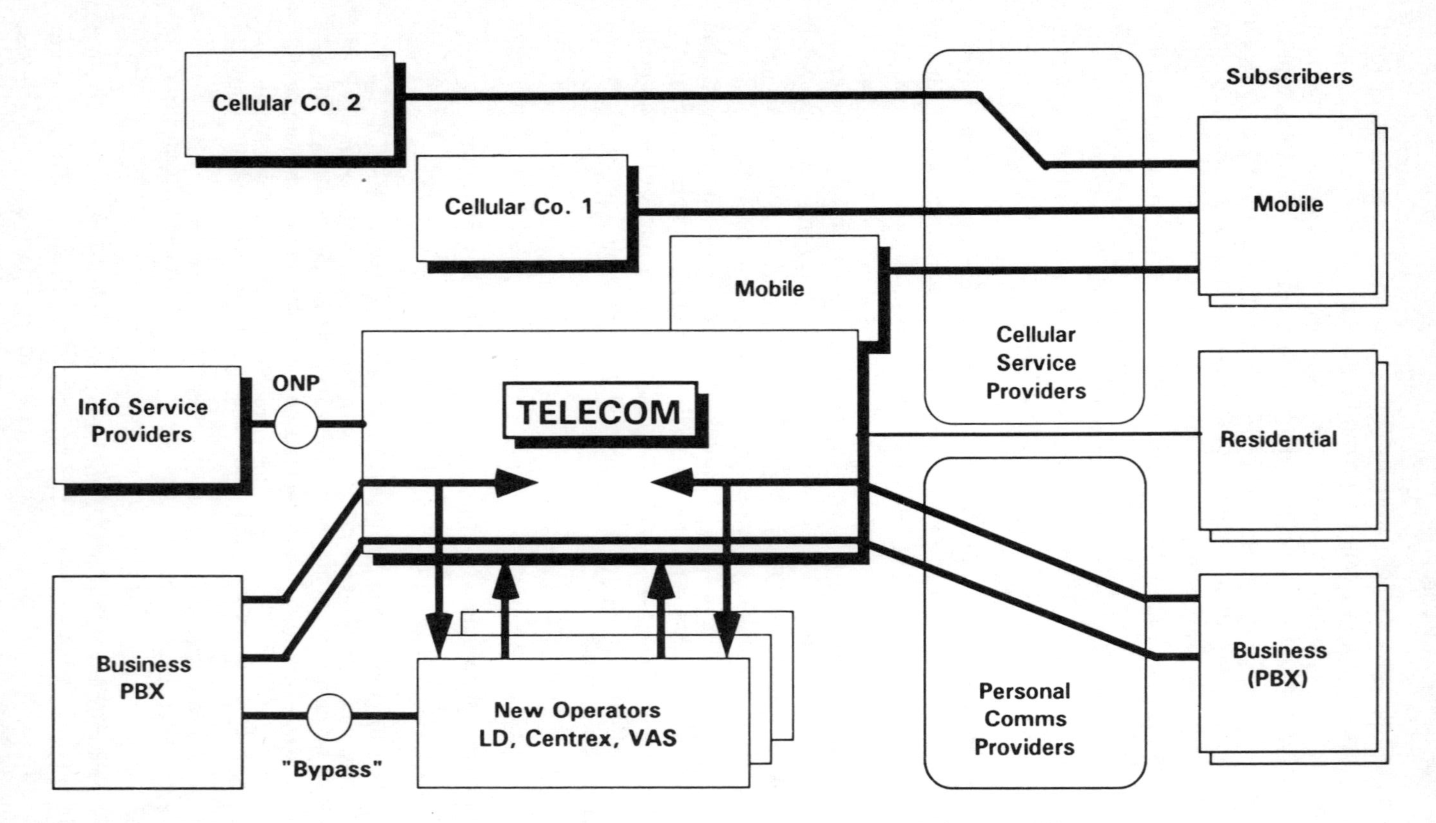


Figure 8.1 *Telecommunication operating environment, 1990s (Ericsson, interview, February 1990)*

For example, 'computer systems would provide services, and communications networks would provide access. This was the idea behind Bellcore's Intelligent Network-1' (Soderberg 1991).

Ericsson had considered the design of the intelligent network as early as 1985. The company's forays into this area began in 1986 when it joined forces with DEC to develop information systems for the banking sector. Ericsson had also co-operated with Honeywell in the adaptation of the MD110 PABX for the American market. The company had launched a study with IBM in 1987 on the way in which co-operation could be established between a public telecommunication operator and a data-processing vendor. The aim was to develop a design that would be close to the concept proposed by Bellcore.[8]

Technical solutions for advanced functions were to be developed to give customers all the benefits of a private network. Virtual private networks, for example, would rely on databases listing the required service features for each customer. Other services were expected to include freephone and credit card authorization, but there were no plans to develop joint products.

The Ericsson view of the evolution of the intelligent network is encapsulated in the following explanation offered by an interviewee. The problems in developing operational network exchanges are those of capacity rather than of quality. When automatic telephony was introduced, personal operator services disappeared and, for a century or so, engineers have been attempting to build back into the network the intelligence that an operator provided. The real breakthrough came with computer or stored program control (SPC) switching, when computers began to provide an approximation of the human brain without any loss of capacity.

Combined with the arrival of CCSS7 and computerized switching software, service functionality could be separated from the switch. Ericsson regarded the process of intelligent network development as one of unbundling the service software from the standard software in the switching nodes. The new architecture would eliminate the need for node-by-node installation of new services.

Ericsson expected that the intelligent network would bring a blurring of the accepted demarcation lines between public and private networks. In effect, the design concept was a mirror image of the competitive environment facing the manufacturer's customers in both the public switch market and the PBX market. The ideal was for customers to be able to pick and choose from a variety of private and public network communication infrastructures that would interwork to provide the best balance of services. The functional modularity of the AXE switch would provide the basis for a virtual network or 'network within a network'. Business customers could be allocated transmission routes as if they were part of a private network. The AXE switch would also provide digital leased line services which would allow end-to-end transmission routes (Thorn Ericsson 1987).

Put another way, the Ericsson model of the intelligent network would use the AXE modularity to support customized service applications with complete control for the network operator or service provider. Ericsson would offer the platform, and network operators would define the services offered to end-users (Ericsson Ltd 1990).[9] It was recognized, however, that the ideal technical balance would be skewed by standards, prices and the interests of the suppliers in the market.

The AXE allows intelligent network functionality to be distributed across a network, not only in centralized Service Control Points but also in local, mobile and transit nodes as Service Switching and Control Points (SSCPs). As Ericsson put it, 'the cardinal rule when designing today's networks must be to ensure that they are designed for change and evolution' (Ericsson Telecom 1990). The Ericsson concept for the intelligent network had taken account of the fact that network flexibility needed to be implemented with analogue and digital technologies, and with circuit or packet switching techniques.

Ericsson's perspective illustrated the tension inherent in the technical perspective on network design. Once service logic is separated from the physical network, the software or intelligence can be centralized or distributed through the network. In the original Bellcore architecture, a centralized (regional) Service Control Point was envisaged. The Service Control Point was to be linked through the Service Transfer Point via CCSS7, making the introduction of services rapid and simple. This architecture would have succeeded in lowering administration costs but it would also have slowed the introduction of new services since CCSS7 would need to have been installed. There would have been longer post-dialling delays for some services and functionality would have been limited. Ericsson considered that these design characteristics might be unacceptable from a commercial and marketing point of view.

Alternative distributed architectures could be developed in several ways. Common to these was the fact that the Service Control Point functions could be provided by existing upgraded switches. For example, existing Service Switching Points could be upgraded to Service Switching *and* Control Points in order to offer virtual private network services. Start-up costs would be reduced and there would be greater service functionality and shorter post-dialling delays. The disadvantage would be higher administration costs.

To overcome this problem, Ericsson had developed service platforms which could be deployed in multi-functional AXE intelligent network nodes. New services could be introduced using platforms initially centrally located in the network. As demand for services increased, the functions could be migrated throughout the rest of the network.[10] Software units could be developed by PTOs or by competing service providers. The Ericsson AXE profile of intelligent network services was numerous and included services provided under existing non-intelligent network architectures.[11] The 1990 release of Ericsson's Service Management System was capable of handling only AXE-based network elements. It was expected that this would not change until the relevant standards for network management had been developed (Soderberg 1991).[12]

The Ericsson AXE-based network structure was designed to allow access and transport capabilities to evolve with greater freedom from service control capabilities. But the company cautioned that the ability to separate service software from transport switching should not imply the evolution of entirely separate network functions. Ericsson thought it was simplistic to believe that eventually service execution would be completely centralized in a general-purpose computer or that the telecommunication switches would be depleted of intelligence and become simple transport devices. What was needed was a 'gradual migration of service intelligence from central nodes, under central management control, towards the subscriber access system' (Ericsson Ltd 1988).

The main objective for Ericsson had been to design functional modularity so that all switches would be adaptable to the markets where the company had installed the AXE switch. The intelligent network AXE was being promoted in the United States by 1988.[13] The AXE-10, equivalent to a 5ESS switch using AT&T nomenclature, was brought into service for BellSouth. MCI announced that it would bring three of the switches into service as international gateways (Abel 1988). These Ericsson switches functioned as exchanges and as Service Transfer Points. They could handle customer traffic and signals between the exchanges and the Service Control Points in the network using CCSS7 signalling.

For Ericsson, successful design and implementation of the next generation telecommunication infrastructure were deemed to require two ingredients. The first was appropriate software development expertise. According to one source, 'we think the finesse is in the software and what we've come up with is a solution' (Abel 1988). The second was success in the network management area.[14]

Some Ericsson interviewees suggested that intelligent network design was an incremental or continuous process; it was not a revolutionary architectural concept. From this perspective, the early Bellcore models had been intended to bring the computer vendors into the design process to assist in the specification of the Service Control Point. But it was clear that the real-time nature of telecommunication had not been adequately considered at the outset. This was treated as a consistently observable pattern with respect to the development of advanced communication services and was not related specifically to the development of intelligent networks.

In Ericsson's view, the Bellcore proposal to use standard operating systems and to develop software with internal interfaces in the switch had been abandoned because there had been little or no movement towards the use of standard operating systems. The Ericsson architecture used a Service Script Interpreter which was a proprietary system. The company had considered the possibility of creating a multi-vendor open platform, but had decided that this was not advantageous because the software would have to undergo costly modifications to integrate the products of different manufacturers.

The idea that it would be possible to let any service supplier gain access to the public network was regarded as a dream by Ericsson interviewees. In this company's view, there would need to be a public network provider who could take control and responsibility for the implementation of services. Third-party service vendors or customers would build their services under the public network operator's control. This was a necessity to prevent the technical failure of the network.

Ericsson representatives noted that the key problem would be standardization. In Europe, they sugggested that too much concern had been focused on standards for interfaces on the internal side of the network. Ericsson managers argued that on the internal side of the network, the equipment manufacturers can exercise their competitive strengths. The ability to design a network was the source of their relative competitive strengths. Standards would only be needed externally to the switch. Standardized systems for network management would also be extremely difficult to achieve since they were at the core of the public network operator's competitive advantage. Interfaces between the Service Switching Point and the Service Control Point could be standardized to enable third-party access, but those within the Service Switching Point should not be. Ericsson observed that technical standards for the network were beginning to become abstracted from the realities of the marketplace. It was in the rivalry between switch manufacturers and the positioning of PTOs that the design and standards for the intelligent network would take shape.

Ericsson suggested that the intelligent network architecture was simply a reflection of the PTOs' need to generate traffic in the public network. More efficient routing techniques could be used if public and private network traffic were integrated more closely. Ericsson interviewees argued that, even in the 1990s, many PTOs had not seen the shift that was taking place in the market. They still saw their customers as subscribers to a set of clearly defined services. These continued, for the most part, to be defined in terms of optimal performance characteristics rather than in terms of their responsiveness to what customers seemed to require.

### Dominant Network Designs and Strategies

The intelligent network for Swedish Telecom was simply part of a long-term strategy of introducing more advanced revenue-generating services into the public and the private network. The company wished to do this without risk to existing investment in the network and without the need to move away from the relationship with Ericsson which had emerged over the years. The modular design of service upgrades made it possible for Swedish Telecom to move incrementally to satisfy the needs of its largest corporate customers and to bring selected premium services into the public network.

The Swedish Telecom engineering perspective on the intelligent network was the most realistic of all the cases examined in this study. Its managers responsible for network design were concerned about how new technical configurations could be integrated with existing functionality, and whether revenue-generating capability would be associated with new services. The marketing division of the company was also very sceptical of a rapid take-off of intelligent services in the consumer market.

Ericsson believed that the original Bellcore intelligent network design had been misconceived. The early architecture had treated the switches as unintelligent nodes and proposed to move all the functionality, except number recognition, into a single central node, the Service Control Point database. Off-line management systems were to be used to program the database. This would have created a service specific intelligent network and enabled the RBOCs to introduce a freephone service without breaking the terms of the divestiture agreement.

The later Bellcore architecture proposals were believed to be closer conceptually to the Ericsson approach.[15] Abandoned in the United States because they required the full introduction of CCSS7 and updating of many local exchanges, an interim architecture was produced instead.[16] Ericsson saw this as a compromise between Bellcore and the RBOCs who were then unable to agree among themselves. Following this failure to reach agreement, the manufacturers began working toward interim proprietary solutions and standards, and each of the RBOCs began implementing intelligent networks as a reflection of its own commercial priorities.

Similarly, in Europe, no common standard for intelligent network services such as freephone, virtual private networks, personal numbering systems or televoting had emerged. Ericsson believed that there were unlikely to be standards in the immediate future and that each of the European PTOs and the manufacturers would pursue their own commercial priorities, just as was happening in the United States.

## Notes

1 Interview with Swedish Telecom, 27 February 1990.

2 A Centrex service takes a number of subscriber lines and translates incoming numbers into actual subscriber numbers.

3 Interview with Swedish Telecom, 27 February 1990.

4 The challenge to reduce production costs was visible in moves by Teli, a subsidiary of Swedish Telecom and Ericsson, to transfer telephone set production to the Far East.

5 The year 1989 saw changes in the organization of the Swedish telecommunication equipment market. Swedish Telecom moved to acquire the Ericsson sales company which operated within the Swedish market. Until 1989, Ericsson had marketed all products developed by the jointly owned Ellemtel on the external market, while Swedish Telecom marketed them domestically. Ericsson's attempt to set up its own domestic sales office led to objections by Swedish Telecom.

6 For example, in 1990 the company was experimenting with a database provided by the Netherlands PTO.

7 Interview with Swedish Telecom, 27 February 1990.

8 Bellcore's IN/2 version, with service independence from the physical structure of the network.

9 Granstrand and Sjolander (1990) have argued that Ericsson's devotion to the AXE with its

modularized software which contributed to its advances in developing the intelligent network architecture was established through a strategic decision in the early 1970s. The company opted for flexibility and modularity of a fully digitalized system in preference to a centralized telephone exchange, the AKE, which was under development at the same time (Granstrand and Sigurdson 1985).

10 The Ericsson approach included the implementation of a Service Script Interpreter and the TCAP (Transaction Capabilities Application Part), both of which are located in the AXE central processor. Service Script Interpretation is a family of service-independent Feature Modules that can be linked in any logical sequence to provide a service, e.g. time-of-day, call gap, queuing, screening, business customer identification, billing number management, etc.

11 Services include freephone/advanced freephone, universal number, emergency numbers, partly paid calls, mass call attempts, information delivery services, information retrieval services, access verifications, incoming-call screening, outgoing-call screening, simple virtual networks, customer control of own data, statistical reports to customer premises, customer control with terminal, credit card calling, personal numbers, virtual private networks, nation-wide Centrex, call-forwarding, not line, queuing, etc. (Ericsson Telecom 1990).

12 Standards were being developed by the International Telecommunication Union's CCITT, the European Telecommunication Standards Institute and the T1 committee in the United States. The American National Standards Institute (ANSI) is an accrediting body which accredits committees such as the T1 committee which plays an important role in setting standards for telecommunication in North America.

13 There was speculation that the development costs for the American market had amounted to some US$1 billion. This was refuted by company officials who suggested a figure one-third that amount. Nevertheless, to adapt the switch to the American market, Ericsson had established a research and support centre in Texas with more than 200 programmers and engineers.

14 Interview with Ericsson, 26 February 1990.

15 The Bellcore IN/2 version, see note 8.

16 The Bellcore IN+1 version, an interim stage before full deployment of Bellcore IN/2.

# 9

# Collaborating with Rivals in Telecommunication

> More serious is the habit of making a model that pretends to represent reality, though in a highly simplified form, deducing conclusions from it and then using it to recommend policy, without first checking on how far the simplified assumptions correspond to the situation in which policy will operate.
>
> (Robinson and Eatwell 1973: 11)

## Rivalry via Competition and Collaboration

There are two main ways in which intelligent service applications can be introduced into public networks. The first is to implement applications in switches that are decentralized throughout the network. In this case, the software is generally developed by traditional switch manufacturers, frequently using non-standardized interfaces and software. The risk to the PTOs is that investment will lead to increasing incompatibilities and complexity in the management and operation of the network. The second way is to introduce intelligent service applications via the installation of centralized databases which are physically separated from the switch. Centralized databases serve as a platform for the provision of services over fixed or mobile networks. If compatible standards emerge, then the PTOs will be able to co-ordinate components offered by multiple suppliers (Onians 1989).

In this technical environment, traditional switch manufacturers can attempt to defend their market positions either by building additional core expertise in computing or by embedding intelligence within the switch rather than in centralized databases. Conversely, computer manufacturers and software houses tend to argue in favour of the use of centralized computer components. This would avoid the need to move their businesses into the relatively unknown territory of real-time switching technology. Figure 9.1 graphically represents the range of strengths of different types of manufacturers. It is clear that IBM and DEC are seeking to concentrate their core competencies in the Service Management System and Service Control Point components of the intelligent network, both of which involve substantial data processing. In turn, the telecommunication switch manufacturers are moving into the computing field from their basic competencies in the production of Service Switching Points and Service Transfer Points.

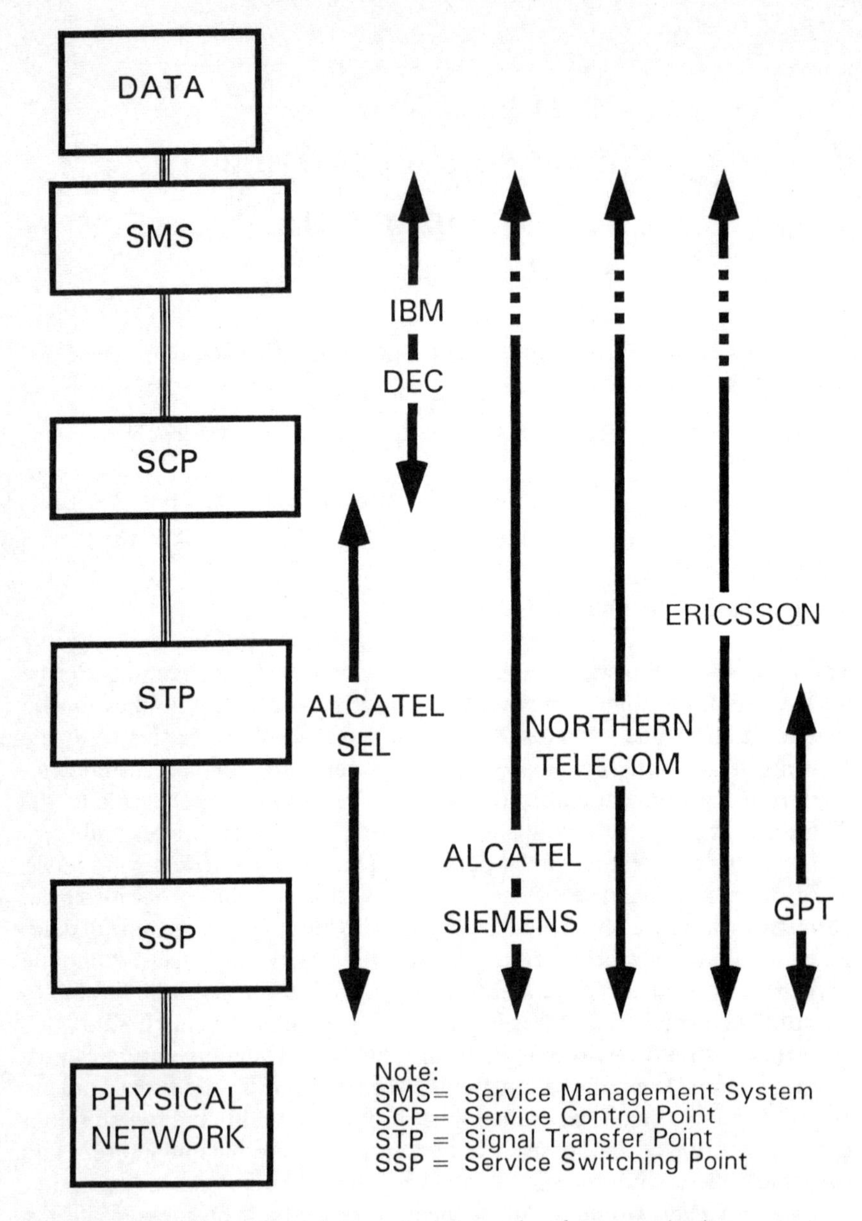


Figure 9.1 *Intelligent network strategic options: the telecommunication manufacturers*

The design opportunities and constraints inherent in the intelligent network architecture threaten to break down the traditional vertical and quasi-vertical relationships between the PTOs and their equipment suppliers. However, whether new suppliers win a growing share of intelligent network and related markets depends less on the technical superiority of alternative approaches than on the strategies adopted by PTOs and the manufacturers. Today, the European PTOs can purchase future generations of equipment from foreign-owned companies such as AT&T, Northern Telecom, DEC or IBM. Alternatively, they could wait for European-owned manufacturers to develop intelligent network products that fit more closely with the historical design characteristics of their networks. This chapter focuses on the rivalry among the European PTOs and the manufacturers and on the way in which this is shaping choices about when to compete and when to collaborate.

**An Industry in Transition – Expected Strategies**

The choice of a competitive or a collaborative strategy depends upon a number of factors. First, innovations in technical design including the advent of the intelligent network architecture encourage new entrants in the market. Secondly, shorter product life cycles and escalating R&D costs have forced telecommunication manufacturers to seek new ways of ensuring that their products find a market. Thirdly, liberalization in the United States, the United Kingdom and Japan has begun to force open markets. In these countries, although vertical and quasi-vertical relationships between the PTOs and telecommunication manufacturers remain intact to a degree, multi-sourcing is increasingly common. Also increasingly common are various forms of co-operation between users and/or suppliers which are motivated by the scale of the required investment in advanced networks and applications. In other words, this type of collaboration grows out of competitive rivalry.

As Schumpeter observed, competition is endogenous to capitalism; it cannot be eliminated (Schumpeter 1943). Collaboration lives alongside competition, giving rise to antagonisms, instability and conflict (Chesnais 1988; Hamel et al. 1989). Collaboration can be organized as a zero-sum game in which one partner loses at the expense of the other. It can also be organized as a positive-sum game for those who are partners in a collaboration at the expense of rival companies who lose as a result of their exclusion.

The modern corporation was created in the late nineteenth century when managerial hierarchies began to be substituted for market transactions. What is arguably new since the 1970s is the way in which firms appropriate scientific and technical knowledge through interfirm agreements which differ from the horizontal and vertical integration strategies of the past. Miles and Snow (1986) have suggested that new forms of collaboration are used to generate competitive advantage under conditions of substantial uncertainty, which is both cause and consequence of the intensified oligopolistic rivalry that has emerged over the past two decades.

The balance between collaboration and competition in the telecommunication industry must be considered in the light of wider trends in the economy, and it cannot be attributed simply to the convergence of telecommunication and computing technologies. Both collaboration and competition are very clearly in evidence in the strategies of firms seeking to build and to operate the telecommunication networks for the coming decades. The question is whether collaboration based on rivalry in the 1990s will produce the market characteristics of the Idealist model or whether it will adhere to the Strategic model. If the latter is the case, the outcome must be examined to determine how the technical system is shaped in the image of the dominant players.

Collaboration has existed in the form of joint ventures and cross-licensing since the 1920s. Contemporary collaborations are much more varied than in the past. For instance, new legally independent and autonomously managed enterprises are often established by two or more parent companies conducting specialized activities to meet common interests. Collaboration can take place between firms of approximately equal strength, or it can involve firms with unequal market share and financial strength. In the 1990s, firms are likely to grow either by internalizing activities previously undertaken by other firms or by extending a web of external linkages and looser ties to form the so-called 'postmodern' organization or the network firm (Antonelli 1988; Chandler 1962; Horowitz 1989; Pisano et al. 1988; Thorelli 1986).

Several motives for the growth of interfirm technical co-operation are common to all R&D intensive sectors. In the first place, there is a need to recover high development costs, and collaboration occurs in highly concentrated industries where competition is waged among a relatively few large rivals. Secondly, with the increasingly rapid pace of technical change in some sectors in the 1970s and 1980s, collaboration is seen as a way to share the high costs of innovative product development. Thirdly, technical collaboration can provide access to knowledge and techniques which can shorten product delivery times. Political and economic power play a key role in competitive rivalry and in collaboration. The decision to enter alliances involves calculating the strength of participants and the way it is likely to change over time. 'Firms form partners for the dance today but, when the music stops, they can change them. In these conditions competition is still at work even if it has changed its mode of operation' (Richardson 1972: 896).

Collaboration can create opportunities for learning and innovation within systems of asymmetrical competitive relations. Effective cycles of learning often require partners to relinquish, rather than increase, control over certain activities. Decisions about whether to internalize technical activities such as software development, to purchase switching components on the market, or to engage in collaborative development projects, all occur in an unstable environment. Under the Idealist model of competition, the appropriate governance structure is determined, in part, by the costs of managing transactions under alternative structural and organizational forms (Williamson 1975; Williamson 1985). Transactions will tend to be internalized for technologies and services that are very complex, whereas market exchange will tend to predominate when products are relatively simple and highly standardized. An intermediate alternative between internalized transactions within a corporate structure and transactions between independent firms on the market is some form of co-operative venture (Ciborra 1992). In the Strategic model, many firms will engage simultaneously in a wide spectrum of internal, external and collaborative activities.

The likelihood of technical and other forms of collaboration that support the commercialization of products depends upon the range of strategies available to a firm when it is confronted by rapid technical change. Changes in technology and the political and economic environment induce firms to try to get ahead, or at least to keep pace with their rivals. If survival depends on the introduction of new products, firms can adopt two strategies: they can introduce a new product design in a bid to achieve technical leadership, or they can avoid initial development and financial risks by emulating the designs of others.[1] Firms tend to specialize in activities where they believe they can gain competitive advantage. They tend to grow and integrate into areas where their internal resources are complementary such that 'activities are complementary when they represent different phases of a process of production and require in some way or another to be co-ordinated' (Richardson 1972: 889).

Telecommunication firms specialize in advanced microelectronics and software-based technologies which exhibit systemic features based upon functionally related component parts. The intelligent network architecture in telecommunication is a good example of this. Production of systemic technologies can demand both a clear division of activities between competitors on the market and a substantial degree of co-ordination. Producers of individual parts of a complex network system must often jointly define standards for compatibility among technical designs developed by collaborating partners. Although, notionally, all these options are open to the firm, the choice will depend on a great deal more than a static assessment of the strategy that will result in reduced transaction costs. The Strategic model suggests that firms will also consider the cyclical processes of technical design and the complex determinants of innovation and strategic positioning in a market characterized by rivalry and uncertainty. These determinants include their assessment of historical strengths and weaknesses as well as their newly acquired competencies.

Among the factors in the Strategic model which contribute to whether a firm will choose to collaborate or compete is the degree to which a technical innovation can be protected. This factor is closely linked with the dynamics of change in the institutional environment; for example, legal treatment of intellectual property rights, regulation of market access and so on. Some technologies are difficult to emulate and fit within a tight appropriability regime, whereas others are virtually impossible to protect. For example, the software embodied in telecommunication switches is subject to a loose appropriability regime because reverse engineering and other techniques can be used to emulate technical designs (Arrow 1962).

Decisions to enter into technical collaboration also depend on the development cycle in an industry, and the extent to which a dominant technical design has emerged in the market (Abernathy and Utterback 1978; Dosi 1982; Nelson and Winter 1982). In a period of rapid change, clear divisions between R&D undertaken by different firms often break down. Innovation calls for the exploitation of synergies and cross-fertilization of knowledge among scientific and technical specializations. Dosi has suggested that technology development cycles are divided into two periods by a dominant design watershed (Dosi 1982).

In the first cycle, competition is based on the creation and modification of designs. This involves 'learning by using', in which the experience of utilizing the technology is fed back from the user and embodied in subsequent design modifications (Rosenberg 1982). Commercial success is particularly dependent at this stage on a good understanding of customer requirements (Lundvall 1989). The discussion of telecommunication switch design in the preceding chapters suggests that there is only the weakest of links between the mass consumer's requirements and the design of the functionality of the intelligent network. There are stronger links between the introduction of enhanced public and private network intelligence and the requirements of multinational firms. However, even here, there is little evidence that these requirements are easily translated into the design and standardization processes for the network interface.

There is evidence that, even before the dominant design watershed has been reached for the intelligent network, both PTOs and manufacturers are seeking access to complementary activities in external markets. Risk reduction requires the acquisition of key R&D inputs since there is uncertainty as to which design will predominate (Viesti 1988). Additionally, collaboration in the development of systemic technologies such as telecommunication switching products has a major advantage. This strategy can result in the enforcement of a standard which penalizes firms excluded from the collaboration.

Firms able to control the development of a new dominant design will use the result to begin an erosion of the existing markets of other suppliers based on older technical designs. In response, leading firms may co-operate to defend oligopolistic arrangements against the intrusion of other companies. Where new technologies modify or weaken existing boundaries between industries, protection against international competition may involve 'clusters of technology-related industrial and/or service activities and product groups, organised around vital basic core technologies' (Chesnais 1988: 82). In the telecommunication field today, the core technologies revolve around the architectural design and the implementation of standards for network intelligence.

After a dominant design has emerged, the high development costs must be recovered in the production and sale of a firm's output. The higher the fixed costs of R&D, the greater the share of the market required to achieve low unit costs of production. After product design has stabilized, there is an opportunity to recover capital expenditures. Economies of scale and 'learning by doing' become the main factors determining competitiveness at this stage. Porter and Fuller (1986) have suggested that the internationalization of markets is encouraging firms to enter cross-border alliances. Collaboration can take less time to establish than the development of in-house expertise and the build-up of market share through internal growth. Collaborations also can be dismantled relatively quickly when the time comes to compete on the basis of the technical design which has been established.

Another reason that a firm may enter a collaboration is in order to eliminate the advantage of an industry leader by internalizing key aspects of the latter's skills and knowledge (Chesnais 1988: 86). Much of the knowledge and skills migrating between companies is not covered in the formal terms of agreements. For example:

> Top management puts together strategic alliances and sets the legal parameters for exchange. But what actually gets traded is determined by day-to-day interactions of engineers, marketers, and product developers: who says what to whom, who gets access to what facilities, who sits on what joint committees. The most important deals ('I'll share this with you if you share that with me') may be struck four or five organisational levels below where the deal was signed. Here lurks the greatest risk of unintended transfers of important skills. (Hamel et al. 1989: 136)

Both formal and informal collaborations are highly visible in the telecommunication industry of the 1990s, and there are many instances of such collaboration in the development of the intelligent network (Duysters and Hagedoorn 1992). However, the previous chapters have shown that equally visible is reliance upon proprietary technical implementations and standards able to differentiate between the architectures of PTO networks and the products offered by their suppliers. This points to a continuing cycle of rivalry characteristic of the Strategic model. The current period of uncertainty does not appear to be a transitional phase of innovative technical development which will reach a dominant design phase technically or institutionally, thus enabling the Idealist model to prevail.

## Strategic Positioning of Global Players

In much of the literature on the development of intelligent networks, a fully permeable global telecommunication fabric is envisaged where supplier and buyer power are perfectly attuned to the competitive market. This is the assumed outcome of the Idealist model. This future is conceived in terms of multiple private and public networks and services which mingle together under the control of multiple suppliers and customers (Consultative Committee for International Telegraph and Telephone, International Telecommunication Union 1990; Eske-Crisstensen et al. 1989). However, in the preceding chapters we have seen that many factors are limiting progress in this direction. Among the tactical weapons that the suppliers, including the PTOs and their potential competitors and collaborators, have at their disposal is the creation of interface standards, control over access to intelligent functionality in the public network, and pricing strategies. Implemented alone, or in combination, these factors can be used to maintain or extend the scope of their markets and to strengthen the strategic oligopoly which is emerging from the monopolistic structures of the past. Evidence of these forms of tactical manoeuvring suggests the predominance of the Strategic model in the telecommunication industry of the 1990s.

These tactical weapons can coalesce to form different strategic options. For example, the European PTOs can form co-operative partnerships in order to strengthen co-ordination in the market, or they can opt for competitive rivalry which will tend towards segmentation of the market. They may also adopt strategies which attempt to defend their traditional markets by enhancing the 'national' identity of their domestic networks, or work towards greater coherence in the market by building up strategic resources – that is, access to capital, knowledge and skills, and so on – through a range of competitive and collaborative deals. Each of these options, taken in either isolation or in combination, offers the possibility to the PTOs of retaining control over access to the telecommunication infrastructure while, at the same time, becoming more responsive to some aspects of customer demand. At this aggregate level of analysis, it is virtually impossible to detect how the underlying tactical tools are being used to achieve corporate strategies. Nevertheless, since much of the literature on the emergence of new forms of competitive and collaborative relations is presented in terms of such stylized options, it is instructive to consider how these options are influencing the PTOs', equipment manufacturers' and large telecommunication users' behaviour.

### *Public Telecommunication Operator Strategies*

One strategic option for the European PTOs is to engage in co-operative partnerships with other European PTOs in order to launch new services. This strategy runs the risk of becoming a target for scrutiny by the European Commission with regard to possible anti-competitive practices. This option is unlikely to succeed if only the European PTOs are involved (Thomas 1992). The 1980s have seen a trail of ambitious plans abandoned by European PTOs hoping to offer managed data networks and one-stop-shopping services. This suggests that, on its own, such a strategy is not viable.

Another strategic option for the European PTOs is to enhance and retain their national identities. In a national market, a PTO can standardize early and support advanced services in the domestic market. The operator can follow its customers to the national border and arrange bilateral agreements with operators in other countries to ensure equivalent levels and quality of service. The objective here is to retain the large users and to pre-empt the migration of customers to foreign, third-party suppliers. But a national PTO may still find itself confronted by competitors in more liberal foreign markets even after a bilateral agreement has been signed with the equivalent PTO.

A third option is to view the existing public infrastructure as a strategic resource upon which to build advanced public services as well as advanced services, such as virtual private networks, network management, call identification and freephone. This is a policy of incremental change, and, generally, it is defensive. It involves a process whereby network designs emerge through co-ordination and planning in order to ensure that the widest spectrum of customer needs is met. This option is closest to the traditional monopolistic form of supply.

A fourth option is to adopt an aggressive business strategy involving the segmentation of the market. If agreements are sought with partners to co-ordinate the introduction of advanced services, the aim will be to share risks and reduce costs where design synergies are believed to exist. This strategy may result in more rapid introduction of new services for some customers, but it carries the threat that partners will capture a significant proportion of the value added generated by new services supported by the network infrastructure. In some cases, the PTO may decide to move into international markets independently by promoting the complete liberalization of market entry restrictions in foreign markets, so that it can support the requirements of its largest customers in their global activities.

The advent of the intelligent network makes the choice of strategic positioning by the European PTOs even more critical. New intelligent features could put the PTOs into direct competition with their current collaborators. Figure 9.2 shows the historical starting points and strategic directions of firms in different segments of the information and communication service marketplace. Computer firms such as IBM are strong in international and national markets. They are moving from their core domain in service applications into communication services. Third-party service suppliers such as GEIS Co. and EDS are highly internationalized. They too are moving into communication services. Private systems are run by third parties via outsourcing agreements or by larger telecommunication using companies. These companies depend on the transport infrastructure provided by the PTOs.

Leased lines have been extended throughout the world to support advanced services. Closed user groups and intracorporate applications tend to run far ahead of the services on offer from the PTOs. These private network operators are interconnecting their services with public switched

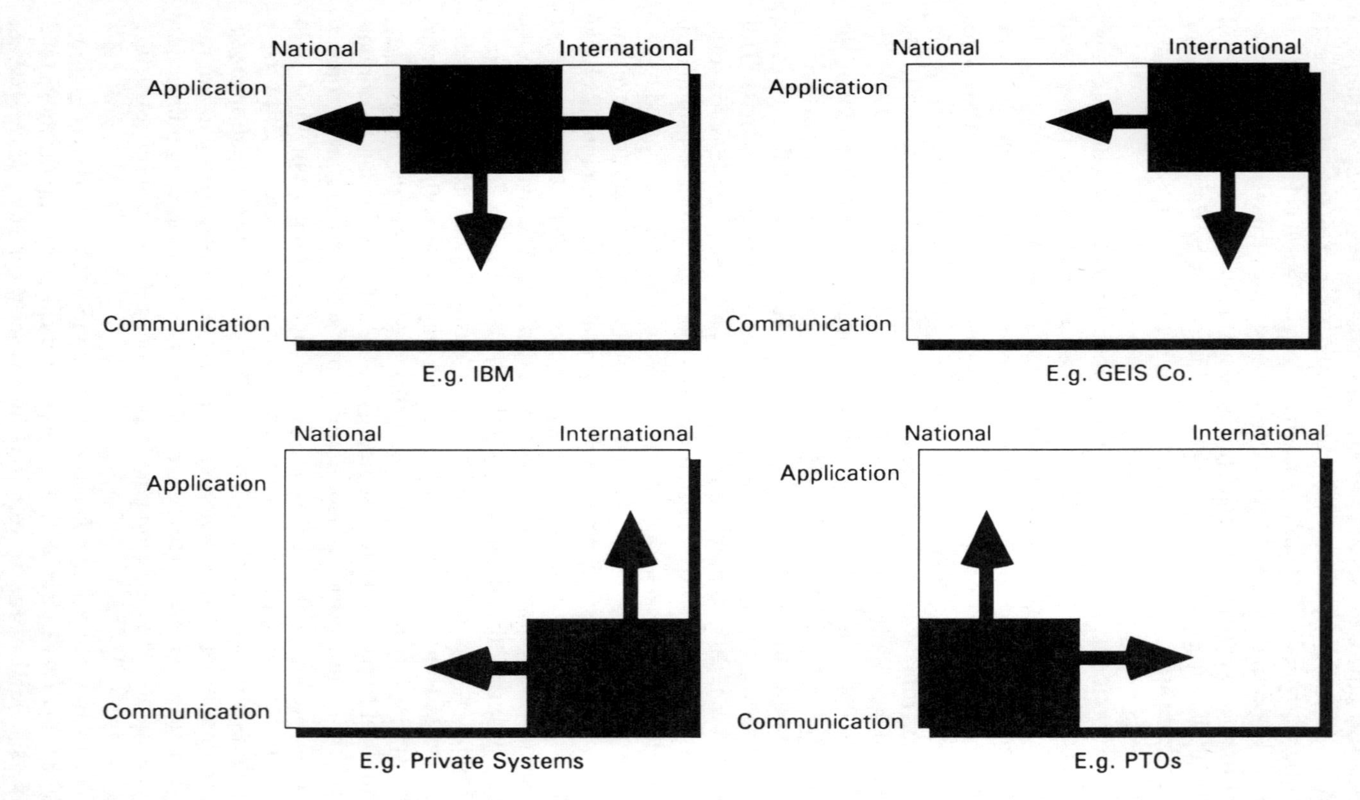


Figure 9.2 *Intelligent network strategic options: service supplier core competencies (after Ericsson, interview, February 1990)*

networks in countries such as the United Kingdom, the United States and Japan, where they can support voice, data and image applications. The PTOs have competence in providing voice telephony, data transmission and a relatively small number of sophisticated services, mainly in their national markets. Relatively few PTOs have experience in supplying all aspects of network management and support for communication services and applications.

The intelligent network is altering the balance of competence among these suppliers. The need for greater co-ordination in the development of software and access to a trained cadre of technical staff able to manage complex global networks is stimulating new kinds of partnerships. These new alliances could give undue advantage to the PTOs whose prospects are strengthened by their large positive cash flows generated by services provided in monopoly or reserved service markets. The strategic choices made by the PTOs in the interest of self-preservation could produce seamless permeable networks, or they could result in network segmentation which is akin to the national segmentation of networks in the past. However, the implications of network segmentation would differ in so far as it results in new patterns of access to the electronic communication environment. This development, which arises out of new forms of strategic oligopolistic rivalry, will have implications for the terms and conditions of participation in the electronic communication environment and will require the attention of public policy.

### *The Telecommunication Manufacturer Strategies*

The technical convergence between the telecommunication and computing sectors had been expected to herald the entry of new suppliers, resulting in greater competition and the emergence of a scenario akin to that envisaged by the Idealist model. The advent of multiple sources of switching and related equipment was expected to allow the PTOs to mix and match the products of any of the large manufacturers who remained in the market after each round of industrial concentration was completed. For adherents to the Idealist model, the development of the intelligent network is an exemplar of one iteration in this process. However, reality is more complex than this. The European telecommunication equipment industry displays examples of co-operative alliances, but also of rivalries which are leading to acquisitions and mergers. These developments are in line with those forecast by the Strategic model. As Arnbak has observed:

> The classical PTT approach strongly influenced the attitude of its suppliers, the manufacturers of public telecommunications equipment. These companies accepted the slow but reliable purchasing cycles and the performance view of telecommunications operators. Not surprisingly, they now have difficulties in entering the much faster computer . . . markets, which emphasise new capabilities much more than high quality of performance. On the other hand, the genuine

> computer companies find it equally difficult to enter the more performance-oriented telecommunications market, whatever their financial strengths. (Arnbak 1987)

That the Idealist model which is premised upon a fully competitive market has not prevailed is due, in part, to the need on the part of network operators to guarantee some degree of standardized compatibility and network integrity. This has led the PTOs to argue in favour of the benefits of establishing long-term contracts with their traditional equipment suppliers. As a result, most markets continue to be relatively closed to newcomers. The intelligent network had been expected to loosen the PTOs' dependence upon a small group of manufacturers. Indeed, it has done so at the margins of equipment procurement in France, Germany and Sweden. In the United Kingdom, there has been a more significant change with the strengthening of GPT's and Ericsson's position in the domestic market,[2] but both manufacturers are designing their products to British Telecom's specifications. This is because it is mainly British Telecom's requirements which form the basis for decisions as to how the public network will be opened to competitors. It is only in the most ambitious versions of the architecture proposed by the equipment manufacturers that signalling and feature applications would be logically separated from transport and switching in the public network. The multi-vendor environments and open networks which could result from this design might begin to emulate the Idealist model as barriers to entry were reduced for firms with competence in telecommunication and computing.

However, the PTOs have shown that they are unwilling to risk completely disturbing historical relationships with telecommunication switch manufacturers. Hence, most collaborations in the telecommunication equipment market have been tentative and unstable. The key technical question being debated is whether the intelligence, the database in the Service Control Point, should reside inside the digital switch or in a separate node. Potential collaborators are faced with a range of options. Switch manufacturers and computer vendors can specialize in providing one component of the intelligent network, construct the entire system or collaborate in the provision of several parts. The battle among suppliers centres on who gains control of the vital Service Control Point – the telecommunication manufacturers, the computer industry, or the PTOs themselves.

In the telecommunication equipment industry, from the mid-1980s onwards, six trading blocs of collaboration among telecommunication equipment manufacturers have emerged on the world scene. These are shown in Figure 9.3 to be coalescing around AT&T, Siemens, Alcatel, Northern Telecom, Ericsson and several Japanese manufacturers.

In the AT&T case, the main building unit is the 5ESS switch. The Siemens set of alliances has to contend with four switch designs (the Siemens EWSD, the GTE switch, the GPT System X and the Stromberg Carlson DCO). Alcatel has to consolidate its switching systems, the E10 and the old ITT 1240. Then there is Ericsson, whose future is pinned on the AXE. Northern Telecom is the fifth major bloc with its DMS SuperNode, and the sixth is the

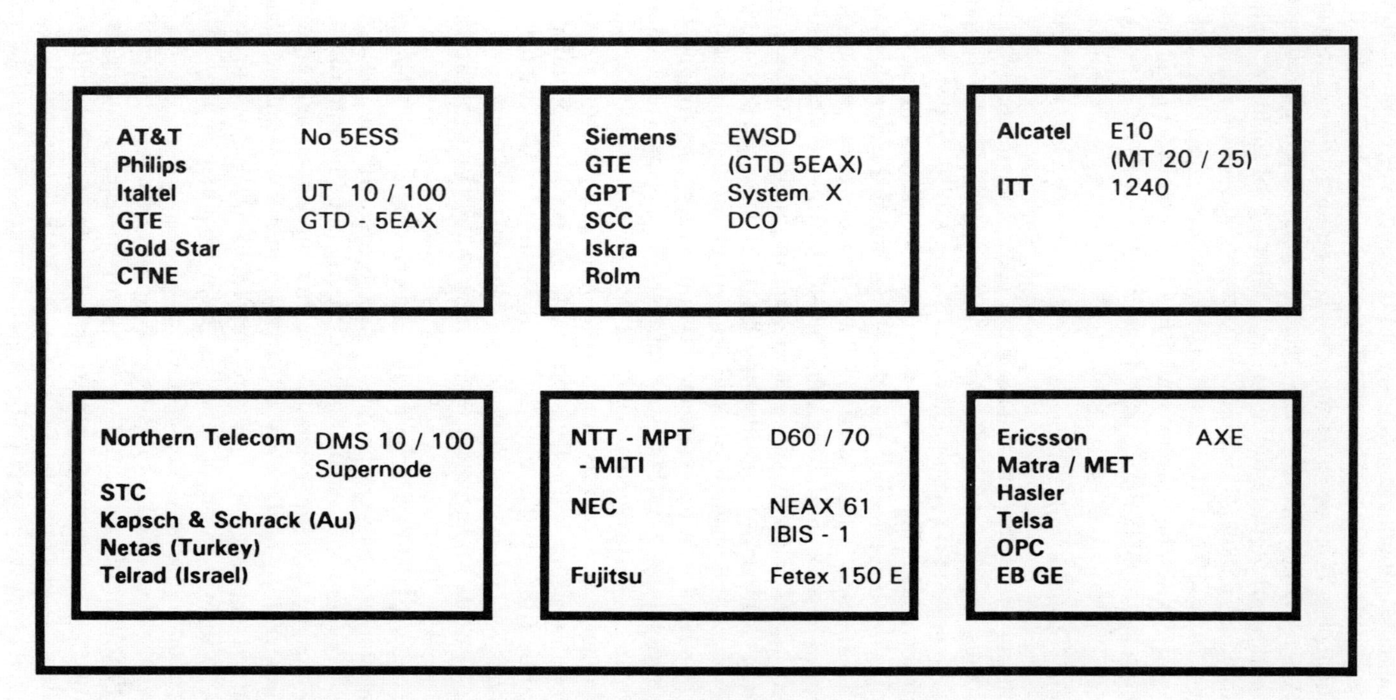


Figure 9.3 *Intelligent network builders (after Ericsson, interview, February 1990)*

switching developments in Japan represented by the activities of Nippon Telegraph and Telephone, the Ministry of Posts and Telecommunications, the Ministry of International Trade and Industry as well as NEC and Fujitsu.

The leading players in each bloc must choose whether to be global players or to look for a market niche. AT&T, Siemens and Alcatel are orientated towards building strength in the market on the basis of their size and by buying into other markets. Ericsson does not see size as the critical factor, and argues, instead, that the costs of the software upgrades and the efficiency with which they can be added to existing networks will be the critical factor determining the shape of these blocs in the 1990s.

*The Computer Manufacturer Strategies*

IBM's ventures into telecommunication are generally considered to date from its acquisition of Satellite Business Systems, shares in MCI, and in Rolm, the American PBX manufacturer, in the early 1980s. However, IBM had been involved in some aspects of the telecommunication industry since as early as 1962 when telecommunication laboratories were established in the United States and France (see Table 9.1).[3] Towards the late 1980s, IBM could claim to have been offering value added network services for more than twenty years. The company was looking to provide customers with comprehensive networking solutions. In 1988, IBM described itself as being involved in the 'total business information system' market, a market that could not be developed without effective communication networking.

IBM's Information Network was constructed to support its internal operations and to provide the backbone for its value-added services.[4] Among the services supported were IBM Managed Network Services, which offered a secure and reliable international network, operated and maintained by IBM. Applications included Electronic Data Interchange, videotex and financial services. Network management was supported by the IBM Open Network Management Architecture and NetView, both proprietary systems.

A key element of IBM's strategy for the 1990s was to capture a share of the market for intelligent network hardware and software. IBM had been active in the Multi-Vendor Interaction task force established by Bellcore in the United States to develop standards for the intelligent network. The company hoped to contribute its mainframe and minicomputers.[5] These were regarded as being appropriate to connect to the telecommunication switch to perform a range of functions from customer inquiry to data communication and network management.

IBM was also interested in providing the software for the intelligent network. In 1987, the company entered agreements to develop intelligent network Feature Nodes, the precursors to the intelligent network nodes. The agreements with telecommunication manufacturers did not generally encompass joint product development or marketing and were aimed at exploiting IBM's competence in data communication and network architecture.

Table 9.1 *IBM's entry into telecommunication*

| Year | Event |
|---|---|
| 1962 | Creation of telecommunication laboratories in Raleigh, North Carolina, and la Gauche, France. |
| 1977 | Launch of Satellite Business Systems. |
| 1982 | Negotiations with Mitel (terminal-equipment vendor). |
| 1983 | Break with Mitel; acquisition of 19% of Rolm. |
| 1984 | Acquisition of COMSAT shares in Satellite Business Systems. Takeover of Rolm. Formation of Trintex (a videotex venture). |
| 1985 | IBM, MCI, Satellite Business Systems agreement. |
| 1986 | IBM, Rolm common presentation of products. |
| 1987 | Joint venture with Network Equipment Technologies, manufacturer of transmission equipment; integration of Rolm into IBM division; intelligent network non-exclusive agreements with Ericsson and Siemens, and Bell Atlantic. |
| 1988 | Stake in MCI sold; 8750 European PBX abandoned; sale of Rolm shares to Siemens. |

*Source:* Systems Dynamics Ltd data in *Communications Week International*, 9 January 1989

A co-operative agreement with Ericsson, for example, aimed to define standard interfaces between IBM data-processing systems and the Ericsson AXE switch. The approach relied upon a centralized control facility rather than distributed intelligence throughout the network. The model would have required a massive database located in the network which would be capable of responding to calls within seconds. The key services were to be virtual private networks, freephone services and credit-card calling. IBM's strategy was to convince PTOs that, rather than wait for enhancements to public telecommunication switches, they could use IBM's computing expertise to develop new services.

IBM increased its promotion of intelligent networks in 1988. Although the company admitted that it had difficulties 'speaking the same language as the carriers', its aim was to convince the PTOs that data processing and greater computer power would give both customers and the PTOs more control over the network. In Europe, IBM set itself the task of selling the concept of intelligent networking to the PTOs. Its strategy was to work with the main public switch manufacturers to develop products. The aim was to 'strike deals with public telecommunications authorities in both value added network offerings and in intelligent networking products'.[6]

It was becoming clear, however, that the convergence of data and voice communication would not result in large incursions by telecommunication and computer manufacturers into one another's markets. An increasingly prevalent view was that 'No supplier can do the whole job. Once a company understands that, it will look for suitable partners' (Korzeniowski 1988a). By 1988, IBM had withdrawn from its telecommunication ventures and had lost considerably in the process.[7]

A trend towards the growth of third-party traffic and shared networks was beginning to emerge in 1988, and companies began disbanding their private networks. This led to IBM's decision to build bridges into the public network by contributing its expertise in network management and software support. The question was how far this strategy could go towards enabling a computing company to become a major supplier of components or systems to the PTOs or to the telecommunication equipment manufacturers. One version of the intelligent network architecture would take the intelligence out of the network completely, leaving the PTOs as *bit transporters*. This would require standardized access points for competing service providers; that is, the intelligent network features would have to be unbundled. IBM saw the European Commission's moves to insist on open network provisioning as one way to meet this requirement. The PTOs would be obliged to make available to users, on a non-discriminatory basis, the capacity to transmit information independently of applications, and tariffs would be required to be cost-related (Armstrong 1987).

In essence, IBM interviewees suggested that the company wanted to repeat its strategy for the American market in Europe. In the United States, the company had taken the view that,

> if I'm an operating company and I can offer a state government statewide centrex with consolidated billing, that puts me in a very interesting position . . . that would be a competitive response to [AT&T's] Tariff 12 and Tariff 16 that would be very attractive . . . if they could integrate a network provider like US Sprint into the situation, it would give them a competitive response to AT&T and fulfil the potential of IBM's NetView. (Sweeney 1989)

The company's view of the intelligent network looked very similar to those views being promoted by the telecommunication equipment manufacturers. For example, as Figure 9.4 shows, in Europe there would be an initial stage of relatively basic private networks followed by intelligent private networks. A third phase would see the interconnection of private networks between customers and suppliers (using the public network), and finally there would be a resurgence of broadband public networks when critical mass had been built up.

IBM was focusing on a 'virtual marketplace' in 1990 which would be served by computing and telecommunication manufacturers who would shrink back to their respective core activities.[8] The company's aim continued to be to remove the intelligence from the telecommunication switches and to place it alongside them where it could be accessed by competing service suppliers. IBM would provide telecommunication transport services only where it was forced to in markets where such services were unavailable, too costly or of poor quality, and would seek to forge alliances with the PTOs.

This would be a difficult strategy, particularly as the company recognized that PTOs such as France Télécom and DBP Telekom would resist. In

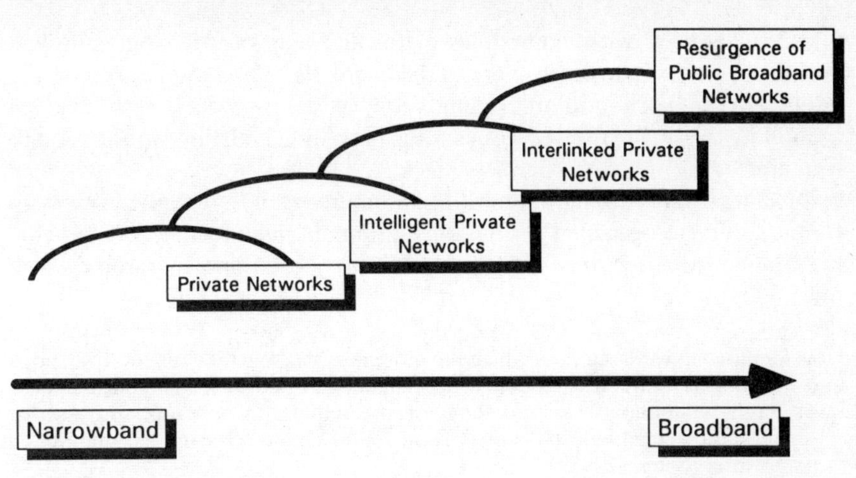


Figure 9.4 *IBM's view of European intelligent network evolution (IBM UK, interview, July 1990)*

France, IBM had not been invited to bid for the first tenders for the intelligent network. France Télécom interviewees suggested that IBM was interested in the intelligent network to obtain knowledge about the internal structure of the public network and its interfaces to strengthen its position as a competitor. France Télécom was regarded by IBM as being very cautious about the core network. IBM felt that France Télécom wanted to 'keep their secrets so that no one of significant size could see the interfaces and compete against them'.[9] From IBM's position in the market the involvement of computing companies in the intelligent network field was dependent upon the design specifications developed by the PTOs. As one interviewee commented, 'If they wanted the computing firms involved they would be.'[10] In Germany, IBM's position was much the same. The issue was one of gaining control over, and access to, the networks operated by the PTOs.[11] IBM regarded the unbundling of service functionality as essential to the growth of competition in the services area. If the regulatory environment could not secure open access, competition would not develop.

DEC had developed a very similar view of the public and private network interworking requirements. For example, one spokesperson believed that,

> by the year 2000 all the Top 500 Fortune companies will have extensive networks of over 50,000 connected workstations. All those networks will then be connected between them through various public services and will provide access to most small companies and eventually to private houses. There will be instant access to all information needed to perform any task. (Giacoletto 1989)

DEC saw the intelligent network as the 'wave of the future'.[12] The company announced a programme for 'enterprise wide networking' in 1989, and formed alliances with companies with the aim of offering complete business solutions to customers. DEC saw the implementation of the intelligent network and interconnection of new services with existing capabilities as a distributed processing problem, and this would require standards.[13]

DEC was undertaking network architecture studies with the American RBOCs, the European PTOs and the equipment manufacturers to develop the optimal architecture. This would be distributed and integrated, such that,

> as we move towards services which are more complex and take longer to set up, a centralised, hierarchical system will be unable to deliver acceptable response times. It will simply take too long for the request to be processed and for sequential commands to be transferred back to the switch even with unlimited bandwidth. (Setchell 1989)

Computer hardware and software would be distributed throughout the network to keep pace with increased traffic and the processing requirements that advanced intelligent network services would generate. In addition, a distributed environment would match the geographically dispersed nature of the existing telecommunication network. PTOs would be able to deploy exactly the right 'increments of computer horsepower wherever they are needed' (Setchell 1989). DEC's study of the configuration of the centralized intelligent network showed that, although it allowed host computers to be accessed over the long-distance network, the result was less reliable than the decentralized model, and it would cost more than connections to a local processor in a distributed network (Schenker 1989). DEC's VAX cluster computers, redundant configurations and networking capabilities were expected to provide the fault-tolerant dependability needed to meet PTOs' requirements.[14]

Like IBM, DEC had entered a variety of non-exclusive agreements with other equipment manufacturers to develop intelligent network products and services. In 1989, the company agreed partnerships with Siemens Public Switching Systems Inc., DSC Communications Corporation, Cincinnati Bell Information Systems and Datap Systems Corporation of the United States to develop call-processing applications.[15]

Other computer manufacturers, such as Tandem Computers, have also entered the intelligent network market. This company manufactures transaction processing systems and is well known for its fault-tolerant computers.[16] Tandem has marketed software for the intelligent network through its Tandem Telecommunications Systems Inc. subsidiary. Applications include freephone, virtual private networks, and calling-card applications as well as communication links. As early as 1986, Tandem had supplied one RBOC's network management software program.[17] The company had announced its Service Control Point, a Service Management System and a service creation environment in 1990 for use in both the public and private networks.

### *The Customers' Virtual Networking Vision*

The level of resources committed by the PTOs and the telecommunication and computing equipment industry to the development of the components of the intelligent network is significant. The design activity is aimed primarily at capturing, and in some cases re-capturing, the large business customers' traffic and to ensuring that these firms do not opt for third-party supply of services or self-supply in a liberalized telecommunication environment.

Many large telecommunication service customers do not believe that the public telecommunication network can ever be upgraded to the point where it meets all their needs (Mansell and Sayers 1992; Norton 1990; Schenker 1991). Most of these customers suggest that there will always be a role for independent or private networks. For these firms, the issue is not simply the degree of complementarity or substitution between public and private networks. The issue is the spectrum of choice within the framework of the total service requirements of the firm and the extent to which functionality is available in the public switched network.[18] Several variables are considered when planning the use of public and/or private networks. Sometimes price is a determining factor but this is moderated by quality, reliability, security and other considerations (Regan 1988). Most large users suggest that virtual private networks which depend on the intelligence embedded in public networks would be used if public network services improved in quality. For larger businesses, the main impetus to install private networks for voice services is to reduce costs, whereas in the data field, the incentive to move to private networks is much more closely related to the quality of network control.

Business associations have been created to champion the business customers' views in the United States and Europe. The first telecommunication user organization in the United States was the International Communications Association, established in the late 1940s. Companies had experienced difficulty in obtaining good service from AT&T. Business users found themselves paying high prices for services and were unable to go to alternative suppliers. The International Telecommunication User Group (INTUG) was created in 1974, mainly by business telecommunication managers in the United Kingdom.[19] Company telecommunication managers were concerned about the restrictions they faced on the ways they could build private networks. Generally, INTUG has pursued two objectives: the benefits of an open market in telecommunication supply and the introduction of service tariffs which are related to costs. Another organization in the United Kingdom which represents the business view is the Telecommunication Managers Association (TMA), established in 1958.[20] At that time, the main concern was the long waiting list for new services. The TMA includes about 30 percent of the 1,000 largest firms in the United Kingdom. The Telecommunication Users Association (TUA) includes smaller companies, and it has tended to focus on tariff issues and public services for smaller users. The TMA is more concerned with the British regulatory regime than the European market as a whole, but an increasing proportion of the budget is spent on European standards work.

In France, there is one main telecommunication user organization with 200 to 300 company members. There is also an organization which is similar to the British Computer Society and has about seventy to 100 members.[21] In Germany, there is one main user group, Deutsche Telecom eV (DTeV), whose membership comprises about 100 to 120 small firms, but it is generally led by a representative of a large telecommunication using firm. ECTUA (the European Council of Telecommunications Users Associations) was formed in 1989 to represent the interests of national user associations in France, the United Kingdom, Belgium, the Netherlands, Spain and Germany. Some 80 percent of INTUG's members are ECTUA members as well.

These business user organizations believe they face difficulties in attempting to influence the European Commission to introduce more rapid liberalization measures.[22] INTUG has suggested that the PTOs have been too influential in developing the detailed rules for opening public network access and that too few measures have been taken to unbundle the intelligent functionality in the public network. INTUG has also been concerned each time there have been attempts to extend the open network access rules to private service providers and competitors to the PTOs. From the perspective of large business customers, open networks must liberate the telecommunication infrastructure from the grasp of the PTOs. They must not encumber competing service suppliers with new rules and requirements.

From the perspective of the British-based large user organizations, the United States was the regulatory model for the United Kingdom, and now the United Kingdom should be the model for Europe. The ideal model would be one whose goal is to maximize the options available from advanced public network-based services and private networks. Virtual private networks utilizing intelligent public network functionality and private networks which incorporate intelligence via PBXs each offer benefits to the larger user. Dedicated private networks and PBXs allow corporate locations to be linked by leased lines, and the customer is able to control the network. Multiplexers permit the integration of data and voice. In the case of the virtual private network, the larger users advocate a tariffing system whereby the customer pays only for services as they are used. The choice between public and private alternatives would depend on the tariffs and the functionality of the public network, which vary from country to country.

Pan-European virtual private networks continue to be regarded by most users as networks of the future. In the United States, these networks have already begun to proliferate. PBXs have been linked through the services of long-distance interexchange carriers. Wide-area Centrex and local Centrex services have been networked with PBXs using advanced signalling systems. In contrast, in Europe, users wishing to transmit data across Europe have needed to build private networks using PTO leased lines or to rely on one of the few independent managed data-network service operators which are American-based or -co-ordinated. Where private capacity has been unavailable or too costly, they have been forced to rely upon public data networks, which they believe are too slow, unreliable and insecure for many advanced applications. The larger telecommunication customers argue that the PTOs have failed to grasp the key uses of telecommunication services to provide businesses with competitive advantage. They observe that services such as Electronic Data Interchange and transaction processing generally are not being offered by PTOs. Instead, they are being offered by organizations such as International Network Services Ltd in the United Kingdom, and by GEIS Co. and IBM, internationally. In Europe in 1989 Infonet and its European partners announced its Enterprise Defined Network Service which was described as an integrated hybrid network solution for global corporate data networks.[23]

The term 'outsourcing' was being applied to facilities management, managed data networks and virtual private networks in Europe by the end of 1991. The movement to develop these services was being led by AT&T, British Telecom, Cable & Wireless, and Sprint in a bid to provide end-to-end services. Both the British Telecom and Cable & Wireless strategies towards providing global virtual private networks relied upon technical approaches involving fully managed digital overlay networks.[24]

Virtual private networks for the international telecommunication market began to be announced in 1987 and the results reverberated on the European market. In December 1987, AT&T and British Telecom announced the implementation of a virtual private network between New York and London.[25] In early 1988, AT&T, British Telecom and the Japanese international carrier Kokusai Denshin Denwa (KDD) formed a consortium. These carriers offered a service which was marketed as allowing customers to approach any of the three PTOs to obtain leased lines and a variety of related services between the three countries.[26]

In the face of this activity, at least one industry observer commented optimistically that a combination of open networks and reduced tariffs for the use of public networks would lead to a decline in the growth of private networks.

> One thought provoking, and perhaps contentious, conclusion . . . is that investment in private international networks, used only by a single company, will largely dry up over the next five years . . . the rebalancing of PTT tariffs will reduce the economic benefits of private networks, while the emergence of large, third party operators and the completion of standards for OSI [Open Systems Interconnect] will create alternative ways of providing specialised services that are not currently offered by PTTs. (Communications Systems International 1988)

The move toward the establishment of international virtual private networks came with the development of proprietary standards such as the British Telecom Digital Private Network Signalling System (DPNSS) and the CCITT Q.931 non-proprietary standard that permitted private network systems to access American networks.[27] In 1990, American virtual private network revenues were estimated at US$475 million. There were none in Europe. By 1994 projections were for US$1.85 billion in the United States and US$281 million in Europe (Hayes 1990).

Part of the incentive for companies which offered publicly accessible value-added networks to seek ways of extending their services internationally had been loss of market share in the American market. In 1989 British Telecom purchased Tymnet and by 1990 was positioning BT Tymnet Inc. in European countries. At the same time, AT&T Istel's Global Messaging Services Ltd was planning to locate network nodes covering approximately 85 to 90 percent of the European Community. The target markets for competition with the PTOs were clearly value-added network services and managed data network services.[28] British Telecom was planning an aggressive pan-European strategy in a bid to pre-empt AT&T and Sprint International from capturing the market.[29] AT&T planned to install at least one node in each of seventeen European countries, and Sprint International had plans to have nodes in fifty-nine cities in sixteen European countries by mid-1992.

In late 1991, British Telecom launched Syncordia, but was unable to form a global network outsourcing consortia as originally planned with DBP Telekom and Japan's NTT. DBP Telekom was said to be reluctant to take on a junior partnership role, and NTT was engaged in a domestic battle with the Ministry of International Trade and Industry over its international aspirations.[30] IBM was believed to be close to agreeing with British Telecom in mid-1992 to supply part of the management of its international network. In the same period, DBP Telekom, France Télécom and MCI began talking about closer collaboration on international managed network services putting the future of Infonet into question.[31] An alliance between PTT Telecom Netherlands BV and Swedish Telecom, Unisource, began operating in late 1992 and was beginning to attract the interest of the Austrian, Belgian, Irish and Swiss PTOs. Unisource was intending to offer pan-European private network services with end-to-end connectivity managed from linked network management centres in the Netherlands and Sweden.

The European PTOs that are seeking to claim a share of global virtual private network markets are doing so on the basis of a variety of switching systems.[32] There is little reason to believe, on the basis of the evidence, that switching platforms will be either compatible or based upon open standards and interfaces. The need for gateways to connect networks in the future will continue to call for international and regional standard-setting activity. There also will be a need for innovative procedures for settling disputes and conflicts that arise as a result of the rivalry among these players.

## Network Closure and Technical Design

The development of intelligent network architectures heralded what many believed to be a new era in the supply of public telecommunication services.

The Idealist model would be nearly at hand as a result of the promotion of policies of liberalization and minimum public intervention in the telecommunication market. However, the reality in the marketplace has the characteristics of the Strategic model. The PTOs in Europe and the United States have been motivated to introduce intelligent networks for several reasons (Onians 1989; Svedberg 1989). First, the intelligent network has allowed them to embrace a strategy – namely, virtual private networks – that will shift traffic back onto the public network and generate new revenues. Secondly, at the heart of the intelligent network architecture is an approach to the design of the public network which may reduce the dependence of PTOs on telecommunication equipment manufacturers. However, thirdly, and more importantly, the intelligent network offers innovative ways of reducing the transparency or ease with which competitors can gain access to public networks. This is a trend which runs counter to the premises of the Idealist model.

The analysis in this study shows that the predominant strategies for creating the intelligent public network emphasize proprietary designs and are reflected in the PTOs' reluctance to unbundle advanced network functionality in a way that would stimulate competition to the degree required by the Idealist model. In fact, the patterns of development described in the preceding chapters can be summarized as a set of commercially motivated strategies for the combination, integration or segmentation of network infrastructures and markets.

First, there is a combination strategy. This has been adopted by DBP Telekom, for example, which has, as its long-term goal, the creation of a centrally planned system combining the ISDN and the intelligent network by the end of the first decade of the twenty-first century. In Germany, as in many other countries, ISDN is based upon national implementation of the relevant CCITT standards. The communication requirements of large corporations are to be accommodated by interconnected narrow and broadband ISDNs, a view which closely parallels the Siemens' perspective on network design priorities. DBP Telekom has created an explicit relationship between the ISDN and the intelligent network. In its view, the ISDN has accelerated the implementation of CCSS7, the technical prerequisite for the intelligent network. Protected until recently from entry in the voice telephone market, the company has not had to confront the widespread development of private networks within its domestic market. From DBP Telekom's perspective, the only way to avoid plant write-offs associated with short-term, non-standardized network development is to plan for the construction of a combined broadband ISDN and intelligent network. This long-term plan requires that substantial excess capacity be built into the public network. This approach carries the risk of substantial losses should demand fall short of projections.

The second approach has been an integration strategy, which characterizes the attitudes of France Télécom and Swedish Telecom. These PTOs seek to ensure that the planned obsolescence of telecommunication network resources is timed to preserve the value of existing plant and equipment. In this strategy, the intelligent network architecture plays a crucial role by enabling integration of private and public networks. The control features of the intelligent network architecture also can be used to integrate existing and newer elements of the public network. These PTOs argue that large corporate users will rely on private networks rather than waiting for the emergence of fully standardized public networks. Both PTOs put greater emphasis than DBP Telekom on the need to respond rapidly to changes in demand.

The France Télécom strategy for the intelligent network reflects a significant shift from the 1980s. During the late 1970s and 1980s, France Télécom pursued projects such as Télétel, the videotex service, for the mass subscriber. From 1990, however, the company's highest priority has been the large business customer. Despite the new emphasis on the larger business customers, France Télécom hopes that, by 1995, intelligent network services will be provided to the mass subscriber, thus generating new traffic and additional revenues. For France Télécom, the intelligent network offers a means of satisfying business demand and of fending off growing competition from private network operators. France Télécom believes that these users need a combination of networks including corporate networks, virtual private networks and third-party services. The intelligent network will be implemented primarily to provide business customers with an integrated hybrid set of services.[33]

The integration strategy in Sweden has in common with France the emphasis on responsiveness to customers and investment aimed at interworking networks. Swedish Telecom is seeking to offer 'total system solutions' which cover all aspects of their national and global communication needs. One of the company's competitive responses is to develop virtual private networks, but there was no strong move to encourage the use of virtual private networks as a substitute for private network investment.

The third strategy has been one of market segmentation. From its base in the United Kingdom, the most liberalized of the four national environments, British Telecom is pursuing such a strategy, which emphasizes the introduction of the most advanced services as rapidly as possible to its major business customers. The approach to the intelligent network in the United Kingdom is distinctively market-led. The intelligent network is regarded as just one of several public network features which can be tailored to respond quickly to business demand. There has been little consideration of the need to plan complementary investment in Service Control or Service Access Architectures. Underlying this perspective is the view that digitalization has encouraged the divergence, rather than the convergence, of network options. A mosaic of networks is believed to be likely to emerge: public and private, fixed and mobile, narrowband copper-based ISDN and broadband fibre-based ISDN.[34]

The main trend in the United Kingdom is for intelligence to become more widely distributed throughout private networks. The intelligent network will simply be used to supplement investment in private networks.[35] Because these perform in much the same way as the public intelligent network, British Telecom argues that there will be little demand for universal intelligent services in the mid-1990s. From British Telecom's perspective, the intelligent network will eventually link public and private telecommunication networks by functioning essentially as a gateway between private and public networks via an open interface. However, the installation of intelligence in private networks is resulting in a large number of proprietary signalling and network management systems.

The combination strategy tends to favour the centralization of intelligence at selected points in the public network, and hence to augment the control of the network by the PTO and the role of computer manufacturers. The segmentation strategy tends to encourage the decentralization of intelligent features to the periphery of networks with the potential for greater access by competing network operators.

The differentiation among strategies in European countries can partly be explained by the extent to which the nationally based PTOs have operated within a planned or market-led regime. This has affected the timing of new capital investment. In Germany, the combination strategy has led to large-scale investment in broadband ISDN and intelligent network technologies and excess capacity which is available to support future demand by residential and business customers. The integration strategy in France and Sweden aims to ensure that the potential of historical and new investment is maximized to provide the best solution to immediate business demand. The segmentation strategy in the United Kingdom aims to gain an advantage for British Telecom, and other British-based operators, by meeting the needs of large corporate users through an array of non-standardized voice and non-voice services. This includes the intelligent network, but does not place it at the centre of the competitive strategy.

The strategies of the PTOs have been complemented by those adopted by the major equipment manufacturers. On the one hand, are companies such as AT&T, Northern Telecom, Ericsson and GPT which have opted to decentralize intelligence through the re-design of switches, and the close links between network intelligence and switching functions have been maintained. On the other are companies such as Siemens and Alcatel, which have developed products adapted to the needs of the French and German PTOs with more centralized approaches. Nevertheless, whatever route has been adopted, the ultimate aim has been to achieve a separation of intelligence from the switching equipment and to move control over the design of the functionality of the network into the hands of the telecommunication equipment manufacturers and the PTOs in the long term.

## Complementary and Divergent Assets – Resolving Conflict

A pattern is beginning to emerge as PTOs develop more competitive tariff structures and build intelligence into their networks to serve their largest customers. Figure 9.5 shows how the traditional bilateral agreements between PTOs are being challenged by joint ventures and alliances which also involve the prospects of competition in one another's markets. As shown in the figure, traditionally the PTOs in different national markets provided services by bilateral agreements to originate and terminate services and to share the revenues and costs using accounting procedures. This arrangement is shown as scenario [1]. However, in the late 1980s and early 1990s some nationally based PTOs have begun to open local sales outlets in foreign markets in order to support their customers. This arrangement is indicated as scenario [2].

In markets in which facilities-based competition is permitted by foreign PTOs, the PTOs and their competitors face a complex set of possibilities. For example, a PTO can open its own local sales outlets and provide facilities in a foreign market. This is shown as scenario [3]. Alternatively, a PTO can collaborate with an operator in a foreign market and provide services under a joint supply agreement, as shown in scenario [4]; for example, the PTT Netherlands – Swedish Telecom initiative, Unisource.

The potential is growing for the emergence of complex competitive and collaborative relationships which combine the preceding sets of relationships. For example, as shown in scenario [5], a company such as British Telecom has opened a sales office, Syncordia in Atlanta, to attract large business customers, and British Telecom is seeking to provide some components of advanced network facilities within the continental United States market. Equally, AT&T is seeking opportunities to open sales outlets – for instance, AT&T Istel – and to establish facilities in Britain and the rest of Europe. To date, AT&T's acquisition of Istel in the United Kingdom has not taken it into direct competition with British Telecom. However, AT&T's application to provide voice and data switched services in direct competition with British Telecom, Mercury and other existing British licensees will, if approved, certainly introduce a direct competitive threat. AT&T applied to the UK Department of Trade and Industry in April 1993 for a licence to offer services within the UK domestic market.

The key actors in this set of collaborative and competitive relationships are changing, and the orientation of the largest PTOs is increasingly towards the international market. Their domestic networks are being modernized to support these activities. The rise of supercarriers such as AT&T and British Telecom is no longer simply a vision of the future. It is a reality. The infrastructure modernization process is being designed to support the economic interests of these supercarriers. There is likely to be a second tier of PTOs in Europe – for example, Belgium's Régie des Télégraphes et Téléphone (now Belgacom) – and even a third tier of smaller national PTOs. However, in this unevenly developing marketplace, the dominant technical designs for public networks are being set largely by the internationally orientated PTOs and their associated equipment manufacturers.

The intelligent network, theoretically, is a way of realizing the open network architecture goals which have been established by regulators and

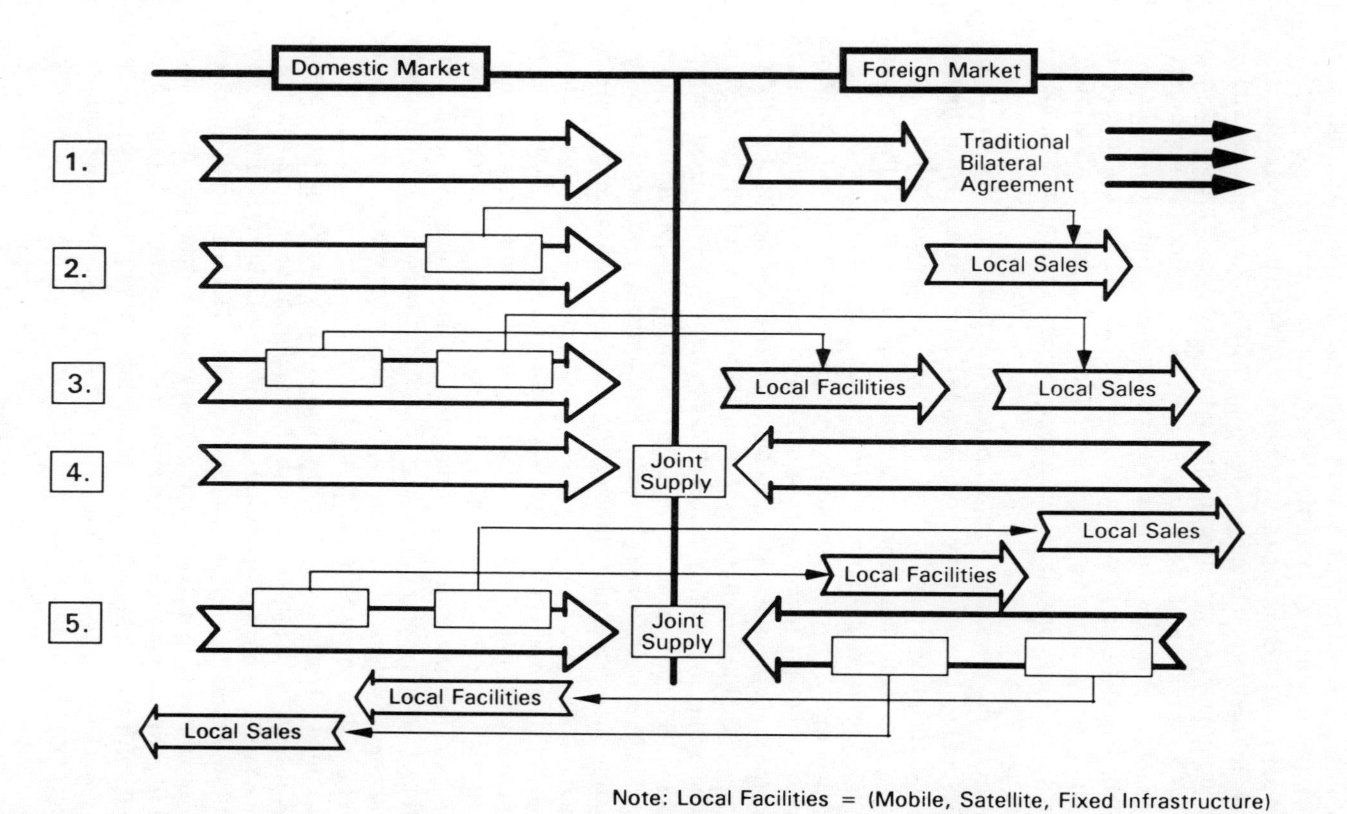


Figure 9.5 *Collaborating with rivals in telecommunication*

policy-makers in Europe and the United States. These are predicated on the availability of standardized network interfaces, unbundled intelligent network functionality and interoperability between network components produced by different suppliers. In both Europe and the United States it has been argued that, in creating a 'level playing field' by opening access and use of the public network, effective competition will emerge. This is the Idealist model. The attraction of the model is clear. If competition can be induced through technical solutions, the demands upon the regulatory apparatus will diminish. The PTOs and manufacturers would also be free to pursue their strategic interests. For those such as British Telecom and AT&T these interests are clearly orientated towards the global market and the largest customers.

However, the standards for the critical components of the intelligent network, the Service Switching Point and the Service Control Point, that are emerging are proprietary, and no single set of standards appears to be emerging as a dominant design in the market. Furthermore, the PTOs in Europe have shown little, if any, willingness to unbundle public intelligent network features for their competitors on the same terms and conditions that they make this functionality available to themselves. In the United States, the implementation of Open Network Architecture by the RBOCs, as we saw in Chapter 3, continues to bring protests by would-be competitors that the result is not creating the 'level playing field'. The intelligent network remains malleable in the hands of the PTOs and their equipment suppliers.

The standardized open interfaces which could give customers and competing suppliers greater control are also those which could erode the strengths of the PTOs. In addition, these open interface standards challenge the distinctive characteristics upon which the telecommunication manufacturers are maintaining a basis for continuing rivalry. These firms continue to seek to lock in PTOs by retaining the major share of the competence required to undertake software modifications and upgrades. Although these traditional manufacturers are being challenged by computing companies and software houses, the computing supplier role appears to be in the periphery of the network, not at its heart.

In summary, two factors are encouraging collaboration among rivals. The first is the systemic characteristic of the intelligent network architecture. Co-ordination is needed to ensure that technical linkages among components are established.[36] Rising fixed costs for R&D related to intelligent network components is encouraging firms to share resources as a way of spreading the risk among several companies. As Ackerman, Vice President of BellSouth has explained: 'With technology moving so fast you do not want to get locked into a fixed position through an acquisition or single partnership. You need to be able to draw on a range of expertise which is often beyond a single company' (Leadbeater 1990).

The second factor encouraging collaboration is a reflection of the customer's requirement for technical interoperability. The PTOs are reluctant to jeopardize relationships with their telecommunication switch manufacturers. They are prepared to bear the responsibility for co-ordinating the development of intelligent network components so as to avoid destructive competition among suppliers. However, as software designs and interfaces become more vital ingredients in the specification of network architectures, so the PTOs are using the re-design of the public network as an opportunity to create entry barriers for their competitors. At the same time, they are maintaining the appearance of open systems and markets through a growing degree of adherence to the principles of competitive tendering and by submitting plans for opening network access to their regulators.

There are a number of external factors which limit the degree to which technical design barriers can be used as an effective defensive strategy. The most fundamental of these is the internationalization of the telecommunication industry. Few manufacturers have acquired the competence to build the entire intelligent network. Even if certain manufacturers do develop symmetrical competencies in telecommunication and computing, the PTOs have become reluctant to be tied to a single supplier because of the pressures to break the vertical relationships of the past. The result is countervailing pressures to produce open standards for network interfaces. Some technical experts argue that the need for interoperability of network components will lead to a technical *détente* based upon a peaceful coexistence among network equipment suppliers and the PTOs (Maher 1989).

This is an evolutionary network environment, technically and institutionally. Only through the deliberate co-operation between PTOs and their suppliers can the intelligent network be constructed in a way that guarantees the interworking of its parts. But competitive rivalry reduces the scope for co-operation. The critical factors influencing the development of the intelligent network are the strategies of the PTOs, the agents primarily responsible for orchestrating the co-ordination of the intelligent network, and their equipment manufacturers.

The fortunes of the major equipment suppliers and PTOs remained uncertain in early 1993. What was not uncertain was that rivalry was resulting in the emergence of new constellations of dominant players in the international market; it was not eroding the sources of substantial market power. The Strategic model has been recreating itself in new forms, and the implementation of open access and distributed control which are the hallmarks of the Idealist model continue to be elusive. This is the environment that confronts the public policy-maker. In the following chapter, we leave the analysis of aggregate strategic orientations to the market, to consider how these strategies are embedded in the choices of intelligent network design configurations. These choices influence the terms and conditions of access to the public network as well as the allocation of cost responsibility to different types of users.

## Notes

1 These strategic alternatives have been developed more fully in the literature on industrial organization. The focus is on incentives which create the conditions for technological innovation and its impact on entry barriers; see Clark 1987; Demsetz 1982; Dowling and Ruefi 1992.

2 By 1992, Ericsson was taking more than a 50 percent share of British Telecom's new orders for digital switching equipment to upgrade local exchanges (Duffy 1992).

3 IBM's telecommunication products included the BSC switch (1964), the PABX 2750 (1972), the PABX 3750 (1975), the PABX 1750 (1980), the Processor 3725 for SNA Network Interconnection (1983), the Token Ring Local Area Network (1985), NetView Network Management system (1987) and the PABX 8750/9750 (1987).

4 The IBM Information Network covered western Europe, Japan and the United States and was accessible from over sixty other countries. The network supports on-line access to IBM mainframes, basic protocol conversion and is moving into managed data network services. The transport portion of the Information Network includes a permanent backbone of 56 kbit/s and T1 transmission links in the United States. In 1989, the Information Network incorporated international value-added service activities that had been marketed as IBM Information Network Services and IBM Business Services. The company drew a distinction between 'network services' or the basic transmission services provided by the PTOs and operators transmitting data from one computer to another, and services which offer several levels of service on top of the transmission service. The Information Network is based on IBM's System Network Architecture (SNA) and SNA Network Interconnect (SNI).

5 These included the 3090 mainframe and the AS/400 minicomputer as well as the System 88, a fault-tolerant, mid-range mainframe computer.

6 Interview with IBM UK Ltd, 16 July 1990.

7 IBM had lost an estimated US$100 million on Rolm, and there had been a 30 percent depreciation in the value of its MCI stock prior to the sale in 1988.

8 Interview with IBM UK Ltd, 16 July 1990.

9 Interview with IBM France, 14 March 1990.

10 Ibid.

11 Interview with IBM Germany, 19 February 1990.

12 The American network was developed by Network Equipment Technologies Inc. Towards the end of the 1980s, DEC was being forced to launch products in response to the increasing threat from IBM in computer networking. DEC announced new co-operative partnerships in its Enterprise Management Architecture programme aimed at network management. The strategy included the development of standardized fibre-optic network systems, interworking equipment, Open Systems Interconnect products to support file transfer, software management, X.400 electronic mail and X.500 directory capabilities. The strategy also included message-handling systems and Electronic Data Interchange software. DEC Europe has a large pan-European E1 backbone network, implemented by the Canadian supplier Newbridge Networks Corporation, with fifty nodes in seven European countries. The EASYnet internal network consists of some 32,000 distributed computers and provides connections to over 700 locations in thirty-three countries. It is used by some 90,000 employees.

13 DEC was active in the American Multi-Vendor Interaction Group, the European Telecommunication Standards Institute Committee addressing intelligent network issues, and the CCITT study group SG-XI on Intelligent Networks.

14 Reliability standards required Bellcore's Service Control Point to have a downtime not exceeding three minutes a year and less than one lost call in a thousand. By 1989, there were over forty-five Digital Service Control Point sites in the RBOCs and independent telephone companies in the United States supporting freephone and credit-card validation services.

15 The non-exclusive agreement between DEC and Siemens in the United States called for product development with a focus on the development of information gateways for PTO networks. DEC's relationship with DSC Communication involved the development of Service Control Point technology. The Service Control Point network featured CCSS7 links and was

aimed at introduction in the American market in 1990. The DEC-Northern Telecom relationship was aimed at enabling PBX voice users to interact with the data capabilities of computers on a real-time basis. In 1987, DEC had announced a Computer Integrated Telephony (CIT) application as a multi-vendor programme to develop smooth software and hardware links between computers and PBXs. CIT was designed with assistance from British Telecom and Northern Telecom. The CIT products allow DEC computers to be linked into private exchanges supplied by the telecommunication equipment manufacturers.

16 Tandem acquired Integrated Technologies, a developer of communication software, and the renamed company became Tandem Telecommunication Systems Inc. (TTSI) in 1988.

17 The Localized Intelligent Network Control for Pacific Bell, one of the operating companies of Pacific Telesis.

18 Private network users continue to use the public network. For example, General Motors is one of the largest users of long-distance telecommunication services in the United States. Only a handful of its thousands of locations originate enough traffic to make it conceivable to construct facilities to provide the location with direct access to the long-distance private network. About one half of GM's 350,000 dial calls in Flint, Michigan, go over local lines; see Huber 1987: Section 3.45.

19 INTUG is governed by an executive committee of six to seven members. The Executive Committee includes American Express, Bank of America, General Electric (US), Digital Equipment Corporation, British Petroleum and Reuters. There are approximately forty associate members, all of whom are multinational corporations with special interests in international telecommunication, including British Petroleum, Reuters, British Airways, Unilever, Citibank, Chase Manhattan, General Electric, Dunn and Bradstreet, IBM, Hewlett-Packard and Motorola. The most direct way in which INTUG exerts influence is by participating in international regulatory and standards-making bodies. INTUG is not permitted to be a member of the International Organization for Standardization since this is confined to national standards-making organizations such as the British Standards Institute. INTUG is an observer member of the International Telecommunication Union CCITT. It has been very active in the European Commission's Directorates General XIII and IV and influences the Business and Industry Advisory Committee (BIAC) of OECD. It is also a member of the International Chamber of Commerce (interviews with INTUG, 15 and 16 August 1989).

20 From 1986 until 1989 TMA grew from about 300 to 700 or 800 members. In 1958, the Government Post Office was taking some eight years to install a public exchange; four years to order and four years to implement. Private networks were constructed in the United Kingdom during the 1950s for voice and telex services: ICI and Unilever had the largest networks, and the large government departments also had private networks. By 1955, the largest private network was owned by the Royal Air Force and the second largest by the War Office. The Admiralty also had a large network. During the years of economic growth in the 1950s and early 1960s, corporations expanded in the United Kingdom and began to decentralize their factories, bank branches, sales outlets, over wider national and international distances. Consequently, large corporations became more demanding in their expectations of the Post Office.

21 These are AFUTT (Association Française des Utilisateurs des Télécommunications et du Téléphone) and CIGREF (Club Informatique des Grandes Entreprises Françaises), respectively.

22 For example, it has been suggested that INTUG has a membership that is biased toward Anglo-American multinational companies. In 1989, 50 percent of INTUG's membership was composed of European multinationals such as Royal Dutch Shell and British Petroleum.

23 MCI had taken a 25 percent share in Infonet in 1990. Shares were also held by Singapore Telecom, Telecom Australia, PTT Netherlands, Régie des Télégraphes et Téléphones, Telefonica, Teleinvest and the Swiss PTT at 5.38 percent and by DBP Telekom and France Télécom at 16.17 percent, and by KDD at 5.0 percent. Plans were also announced for the construction of a global network operating at both the United States T1 (1.544 Mbit/s) and the European E1 (2.048 Mbit/s) speeds as an overlay on the public packet switched network backbone. Infonet switches were co-located with those of the PTOs, thus minimizing the need for gateways and permitting greater network visibility and control. In January 1990, Computer

Sciences Corporation still retained a 30 percent share of Infonet.

24 British Telecom planned to use its CONCERT open systems network management software to control multi-vendor customer networks. CONCERT was designed to be used either by corporate customers on their private networks or by a PTO to track faults. British Telecom promised integrated data, graphics and facsimile at up to T1 (1.544 Mbit/s); voice and video; dynamic adaptive routing and bandwidth allocation, end-to-end network management; interconnection of private networks, support for multiple protocols, packet switched data network X.25, Fibre Distributed Data Interface, open systems Local Area Networks, Transmission Control Protocol/Internet Protocol and Appletalk.

25 In the United States, this was known as International City Centre Service and, in the United Kingdom, as City Direct.

26 The trade press argued that the Account Management Plus service was little more than a new billing arrangement.

27 In April 1989, the PBX problem was not resolved for intra-European Networks. Private network standards had been set by the European Computer Manufacturer's Association (ECMA) and, under a long-standing agreement, approved by the Comité Européen de Normalisation ELECtronique (CENELEC). The ECMA Committee T32 had handled private networking. ECMA had been working on the ISDN signalling protocols that allow different PBXs to interwork on a private network. The standards known collectively as Q-SIG, after the Q.931 recommendation for PBX signalling from the CCITT, describe call set-up and clearing procedures and supplementary services such as call forwarding. The main area of concern was the standards for supplementary services. On the UK side was British Telecom with the DPNSS standard which was developed in 1982 and used on between 2,000 to 3,000 2 Mbit/s lines in the United Kingdom. On the other side were Alcatel and Siemens, which developed a pre-Q.931 inter-PBX signalling protocol called the ISDN PBX Networking Specification. In March 1990, the announcement by Alcatel and Siemens of a joint common specification for inter-PABX signalling using the ISDN D channel pre-empted work at ECMA and CCITT. This created the Inter-PABX Networking Specification (IPNS) with over fifty features, including call-forwarding, call transfer and three-way conferencing. Subsequently, ETSI became involved in a controversy as to whether ETSI standards should apply within PBX environments and this was largely unresolved in January 1993.

28 Taken as a whole, the value-added services market – including credit-card authorization, reservation systems, managed data network services and specialized data networks, public packet switched data X.25 networks, videotex, electronic data interchange and electronic mail – was estimated at a value in Europe of US$5.9 billion in 1992. France Télécom held a 20 percent share of the total; DBP Telekom, 13 percent; Telefonica, 10 percent; Reuters, 7 percent; British Telecom, 4 percent; IBM, 4 percent; SWIFT, 3 percent; Unisource, 3 percent; STET, 3 percent; and GEIS Co., 2 percent; with 'Others' holding 31 percent of the market (Schenker 1993).

29 British Telecom's service is called Global Network Service and runs over the Tymnet network.

30 Syncordia's first customer was Amadeus, the European computerized airline reservation system, which chose the venture to manage its trans-Atlantic link.

31 Under one scenario, Infonet would become a low-end value-added carrier offering messaging, while new consortia would offer more sophisticated services such as facilities management and outsourcing.

32 For example, RTT-Belgacom is basing its virtual private network on Alcatel's System 12; France Télécom's network incorporates the Northern Telecom SL-1 and the Alcatel MF-25 systems. In Sweden the Ericsson AXE 10 is used; and in the United Kingdom, Northern Telecom's DMS-100 is providing the platform (*Data Communications* 1992).

33 In France, in comparison to other European countries, there is the high concentration of large corporations in the Paris metropolitan area where the majority of plants employing over 500 are located. The only other area with a significant concentration of industry is Lyons. A typical firm in France has a private or 'core' network in the Paris area, and uses public network services (e.g. ISDN and Transpac), to connect other plants and offices in the regions. IBM, for

example, has a Metropolitan Area Network in Paris where it has five main sites in the area of La Défense, and these are connected to two other sites in La Gode and Montpellier.

34 This fragmentary approach has been described by Ergas as follows:

> the needs of the largest, most advanced users differ greatly from those of ordinary households, and it makes increasingly little sense to fit both in one box. The logical outcome of this trend is progressive fragmentation of both demand and supply, as niche markets attract new entrants who in turn discover distinct market segments which were previously unexploited. (Ergas 1988: 27)

35 There is concern in British Telecom that distributed intelligence in Delta Networks may be incompatible with the hierarchical functionality associated with the intelligent network.

36 Collaboration over the intelligent network has not reached a stage where it is possible to offer more than a speculative account of the implications of these arrangements. For a discussion of this type of co-operation, see Ciborra 1992.

# 10

# Intelligence and Flexibility for Whom?

> Insofar as social analysis merely replicates . . . a perception of reality without penetrating beneath the apparent technological necessity, it further contributes to the general mystification and reinforces the particular social relations which are thereby obscured . . . modern technology has remained a phantom, a conveniently vague device for explaining historical developments by explaining them away.
>
> (Noble 1977: xxvi, xviii)

## Competition, Monopolization and Rivalry

The design parameters that are shaping the intelligent network have been the main focus of analysis in the preceding chapters. The forces contributing to the evolution of the intelligent network in France, Germany, Sweden, the United Kingdom and the United States are complex and they include both technical and institutional issues. There are those who believe that the technologies which comprise the intelligent network are the harbingers of a new wave of ubiquitous communication services that will be accessible to all. The Idealist model suggests a vision of a fully interactive service environment in which access, via the public network, to all conceivable electronic services is available to customers at ever-declining prices. This evolutionary progression is treated as the result of a technological selection process within a trajectory bounded only by the history of investment strategies, the scope of innovations in telecommunication hardware and software, and the resources available for commercialization. A transitory period of uncertainty during the initial implementation of innovations in telecommunication systems is expected to be followed by the rise of technically optimal public networks. It is anticipated that this evolutionary process will coincide with the structural and operational characteristics of fully competitive markets as envisaged by the Idealist model outlined in Chapter 2.

Some would go so far as to argue that this process of change could be hastened if public intervention, whether in the form of state ownership, regulation or public funding of research and development, were withdrawn. Within the framework of the Idealist model, dominant technical designs and configurations are agreed and diffused in parallel with investment in digital switching and transmission technologies. Institutional adjustments on the supply and demand side of the market, and within the public policy and regulatory environment, create circumstances whereby intense competitive

Table 10.1 *Idealist model assumptions: dominant characteristics*

| |
|---|
| Permeable seamless networks |
| Ubiquity (universal service diffusion) |
| Demand-led telecommunication industry |
| Open systems, common interface standards |
| Co-operative partnerships, transparent network access |
| Minimal regulation to achieve efficiency and equity |

rivalry erodes all significant distorting vestiges of monopoly power in the telecommunication industry.

This view is inconsistent with the trends which characterize the evolution of the intelligent network. The analysis of the intelligent network, as one important facet of telecommunication development, has shown that the assumptions of the Idealist model which are summarized in Table 10.1 generally hold only in specific submarkets and where the customer is a large, globally operating firm. Yet the engineers and technical experts representing the computing and telecommunication industries often forecast the early arrival of the conditions assumed by this model. Although their technical designs for the public network theoretically might support open access to the intelligent network, this is not the predominant trend.

This vision of open network access and competition rests upon a myth which offers the sophisticated products of scientific and technical innovation as solutions to what are essentially institutional problems. In the case of the intelligent network, the myth is that a global ubiquitous network will evolve as a result of its inherent superiority over other solutions. This myth has an important function in providing the rhetoric which is effective in persuading governments to promulgate policies in line with Idealist explanations of the determinants of network development. A myth or commonly held belief often has some very material constituents despite the fact that it is largely without foundation (Smythe 1984:212). The constituents of the intelligent network myth include some very material political and economic determinants, as we have seen in the detailed study of network design. These need to be considered in assessing the merits of the Idealist and the Strategic models as representations of the forces that are shaping the evolution of public telecommunication networks.

Mumford argued some fifty years ago that there was a strong interdependence between technical and institutional change. For example: 'Technics and civilisation as a whole are the result of human choices and aptitudes and strivings, . . . the world of technics is not isolated and self-sustained; it reacts to forces and impulses that come from apparently remote parts of the environment' (Mumford 1934: 6). Recent approaches to the issues raised by the interdependence of technical and institutional change draw attention to the role of institutional factors in the selection of technical configurations.

For example, Dosi has observed that in the process of selecting new technologies, three institutional factors are crucial: the accumulation of knowledge, institutional intervention and 'the selective and focusing effect induced by various forms of *stricto sensu* non-economic interests' (Dosi 1982: 160). However, acknowledgement of the role of institutions in the technical selection process is only a first step. Another is the in-depth analysis of the political and economic institutional constraints of the past and the present and the way these are reflected in the shaping and appropriation of technical change.

For the telecommunication industry, one determinant of change in network design is the incentive to elude the scrutiny of policy-makers. Another is the incentive to protect those technical configurations which provide the basis for competitive advantage in the marketplace. Thus, for the incumbent PTOs, success in the use of standards and design specifications, as well as in advantageous pricing strategies for large telecommunication users, enhances their ability to exploit market power and to create barriers to entry aimed at protecting or extending market share. This same approach is adopted by the telecommunication equipment manufacturers in so far as proprietary interface standards can be used to protect product innovations from competitive challenges.

The PTOs and the telecommunication equipment suppliers are often keen to sustain the myth of the ubiquitous, flexible, open public telecommunication network in the context of the Idealist model. However, the reality of public network development bears more of the characteristics of the Strategic model which are summarized in Table 10.2.

Segmented networks, the reduced availability of more advanced services, and weak forces of competition in most telecommunication service submarkets are the main trends which are forecast by this model. These characteristics are the result of pressures created by international rivalry in the marketplace and the processes of monopolization that were considered in Chapter 9. In this chapter, the ways in which these pressures are articulated in network design are examined.

## A Political Economy of Telecommunication

Although it is the Idealist model which elicits excitement in the trade press and among some members of the engineering community whenever new clusters of technical innovations are considered, it is the Strategic model which reflects the institutional conditions under which technical innovations are produced and used. The former model creates the impression that a new network design will benefit all facets of society and the economy. Nevertheless, it is the Strategic model which needs to be considered if there is to be a clear assessment of the determinants of the evolutionary trajectory for the public telecommunication networks.

The technical determinism encapsulated within the logic of the Idealist

Table 10.2 *Strategic model assumptions: dominant characteristics*

| |
|---|
| Fragmented networks |
| Reduced ubiquity in service diffusion |
| Supply-led industry, multinational user pressure |
| Weak stimuli for competition in most submarkets |
| Monopolization and rivalry, non-transparent network access |
| Increasing regulation |

model has played an important role in the post-war years in encouraging a fascination with the information technology revolution. For example, when Bell first forecast the coming of the post-industrial society in 1959, he looked to information or knowledge as the organizing framework around which other institutions would coalesce (Bell 1973). In the 1970s, Porat called attention to the fact that the information sector in the United States comprised only 10 percent of the labour force in 1900. By the late 1970s, information-related activities accounted for over 50 percent of the labour force (Porat and Rubin 1977). Although there was considerable debate over the validity of the categories used to account for information-related activities, forecasters continued to see information commodities and information technology as the driving forces behind growth and expansion in the international economy.

The burgeoning of information services created a furore over the benefits of the information economy (de Sola Pool 1990). Much attention focused on the introduction of information technology policies, frequently with little regard to the economic, political and institutional factors that would militiate against the fulfilment of policy objectives. Furthermore, the public and private uses of information, and divergent interests in its production, were hardly considered. The main priority was simply to stimulate the production and diffusion of hardware and software.

In the late 1980s, it was becoming clear to many observers that neither information technologies themselves, nor the information and communication services they supported, could be expected automatically to eliminate disparities in market power. Nevertheless, it was possible to find business consultants advocating that successful strategies for competitiveness should rest mainly upon technical solutions such as open network access to electronic modes of communication. For example, a Price Waterhouse *Annual Report on Information Technology* advised business managers to 'grab hold of electronic means to bridge the gap between the receipt of information and the evidently more powerful applications achieved by many Japanese competitors' (Price Waterhouse 1991).

Another technically driven vision based upon the myth of *globalization* is becoming prevalent in the 1990s. Its proponents argue that the process of globalization will, itself, produce a gradual balancing and redistribution of wealth-generating enterprises within the triad of Europe, the United States and Japan. Globalization scenarios link the fortunes of globally operating firms to the availability and use of a flexible telecommunication system that connects suppliers and customers in local and distant markets (Ohmae 1990). Theoretically, no corner of the world will be without access via pervasive telecommunication networks to the products of an interdependent social, cultural, economic and political order. The assumed benefits of globalization have been challenged by recourse to empirical evidence on the actual structure, operation and geographical dispersion of productive activities within global markets (Dosi et al. 1990). The notion that the development of global telecommunication infrastructures and services will ultimately eliminate biases and distortions in the market must be questioned as well on the basis of empirical evidence.

As Noble has argued, analyses of the dynamics of political and economic change that link technical innovation with the automatic adjustment of markets have a tendency to perpetuate the mystification of social, economic and political relations that are embodied in technical innovations (Noble 1977: xxvi). The recent history of European telecommunication and the emergence of advanced telematics systems, such as the intelligent network, can be explained partially by the evolution and diffusion of technical capabilities such as sophisticated signalling, software controlled switching, and so on. But this explanation is of limited value unless it is coupled with the economic and political factors that contribute to the biases operating within the innovation and technical selection processes. These factors shape the design and implementation of operational telecommunication systems. They are tightly enmeshed within the political economy of institutional change and in the strategies adopted by suppliers and some segments of the telecommunication user community.

Noble also observed that simplistic associations between technical change and the restructuring of markets tend to support the status quo (Noble 1977: xxvi). The positions espoused by the incumbents in the telecommunication market when faced with change are evidence of this tendency. For example, the PTOs are seeking new ways of monopolizing the marketplace even as they confront substantial pressures for change as a result of intensified global rivalry. A political economy of telecommunication exposes the power relations that are embedded in processes of technical and institutional change in a way that analysis of co-operative and competitive behaviour in idealized competitive markets does not.

Much research on firm performance in global markets is framed by the rhetoric of the Idealist model. These analyses of technical and institutional change and the learning behaviour of stylized organizations assume away the existence of unequal market power or treat it as a transient phenomenon that will be eliminated by the sheer force of global competition. The Idealist model, in so far as it denies the existence of long-term distortions in the marketplace, simply does not permit the main questions of interest from the perspective of the Strategic model to be raised. In the telecommunication field, these questions concern the determinants of how, and for whom, the intelligent network is being developed.

The process of designing the intelligent network in each of the countries studied in this book displays a technical and institutional selection process that operates within the framework of the Strategic model. The intelligent network is not being designed primarily to encourage the universal diffusion of public telecommunication services. The analysis shows that it is being designed in the light of the particular constellation of economic incentives facing the incumbent PTOs and the traditional telecommunication manufacturers. It is also being shaped by their changing perceptions of the requirements of globally operating firms which are the source of a significant and growing proportion of revenues. Rivalry in an imperfect marketplace dominated by large user demand is leading to a redefinition of the requirements that must be met by the public telecommunication infrastructure.

The evidence in the preceding chapters shows clearly that, in spite of claims that the public network is becoming more open and accessible to users in general, in fact, it is becoming more closely attuned to the needs of a specific segment of the user community; namely, the globally operating firms. The risk, in the absence of effective policy and regulation, is that access to the electronic communication environment for some users will become increasingly limited at the same time that it is being extended for large corporate users. Nevertheless, it could be that restricted access to some aspects of the electronic communication environment is a positive outcome when the requirements of smaller businesses and residential customers are taken into account. This possibility is camouflaged by the rhetoric of the Idealist model. As long as the Strategic model remains in the shadows of policy analysis, there is a strong likelihood that public network design will increasingly accommodate large customer requirements. This will occur without public debate on whether public network capabilities are evolving in tandem with the majority of user needs.

In Chapter 1 it has been argued that some access to the public network infrastructure is essential for participation within all facets of social, political and economic life. However, this does not mean that access to all the most advanced technical systems must be put in place within the public network. In addition, the terms and conditions of such access are essential considerations. Although, it can be argued that the mass telecommunication customer base should bear the costs of sophisticated public networks that are not primarily designed for their use in the interests of industrial policy goals, at very least, this argument should be evaluated for its implications.

### *Network Segmentation and the Strategic Model*

The application of the Strategic model forecasts increasing segmentation in the development of public networks. The contributing factors are as follows. First, segmentation in the terms and conditions of network access is visible in the standards-setting process. Key public network interfaces are being designed to create barriers to entry for competitors to the PTOs. These barriers relate to the levels of access to the intelligent functionality embedded within the PTOs' public networks. They are reflected in the PTOs' reluctance to unbundle intelligent resources to meet the needs of companies who wish to use public network resources as a foundation upon which to offer advanced information and communication services. Barriers are also visible in the tendency of standards-setting organizations, still heavily influenced by the monopolistic PTO engineering culture, to postpone consideration of certain network interface standards until far into the future.

Secondly, segmentation is visible in the pricing strategies being pursued by the PTOs who continue to control access to the most ubiquitous network – the fixed public switched telecommunication network. Speculation as to the degree to which mobile communication services will provide an alternative to the fixed public switched telecommunication network is growing. Services such as Personal Communication Networks and cordless hand sets, when combined with the personalized numbering systems that could be supported by intelligent networks, are expected to challenge the ubiquity of the fixed public network. Although this challenge may be articulated in different ways, it will be shaped by the same forces that are influencing the intelligent network. At present, the penetration of mobile services, even in the western industrialized countries where it is highest, is only a small portion of the fixed telephone network subscriber base.[1]

High cost estimates for access to public network resources for would-be competitors and significant volume discounts for large telecommunication users can shift the cost burden for the design and implementation of the intelligent public network to the bulk of its smaller business and residential customers. These are the same customers for whom the PTOs and the equipment manufacturers openly admit there is weak or absent demand for many intelligent network attributes and for whom most intelligent services are not designed in the first place.

Thirdly, segmentation is apparent in the strategies of the telecommunication equipment manufacturers. They advocate open standards and the benefits of design compatibilities that could emerge in a multi-vendor environment which admits new entrants from the computing and software industries. They also employ designs in the construction of the components of the intelligent network which offer proprietary implementations to their customers. They provide proprietary equipment to support overlay intelligent networks in a bid to stimulate the market for equipment and to strengthen the ability of the PTOs to respond to the large telecommunication users' sophisticated demands. The result is the further segmentation of incompatible networks. While this may create difficulties for the larger users, they have shown historically that they will create pressures for network interoperability in those areas that they believe should be given priority. This entails additional costs which, again, may fall disproportionately upon other users. For smaller businesses and residential customers there is a continuing failure to address their service requirements.

This process of network segmentation is a reflection of the way in which technical innovations in telecommunication are being institutionalized. It mirrors the changing relative market power of major network operators and equipment manufacturers. It is not simply the outcome of technical innovations in telecommunication. The dynamics of monopolization and rivalry, and their expression through the technical and institutional design of the intelligent network, provide an explanation of the strategic changes that are under way. According to Clark, 'Monopolisation *refers not to monopoly as such but to the activities of firms (usually dominant ones) who are seeking to build up/or maintain a position of market power*. Such activities include predatory and exclusionary tactics, for example selling at a loss and exclusive dealing' (Clark 1961: 21; emphasis added). Clark discussed the predatory and exclusionary tactics of interfirm rivalry before software and digital technologies were perceived as creating a technical platform for widespread restructuring in telecommunication. The myth of a global, permeable and seamless public telecommunication network had barely been conceived in the early 1960s, although satellite communication was beginning to emerge as a potential challenge to the terrestrial public network operators.

Technical innovations in telecommunication have contributed new systemic complexities in the provision of telecommunication networks. These are echoed in the complexity of the institutions which now represent the telecommunication supplier, customer and policy communities. Technical and institutional systemic complexity is also reflected in the process of monopolization which is transforming the generic definition of the intelligent network – that is, a flexible, open, physically independent network – into a set of specific design parameters. These parameters embody the political and economic interests of the equipment manufacturers and PTOs. These organizations may establish collaborative, or even collusive, relationships under certain conditions. However, these relationships are developing within the context of international rivalries whereby firms seek to monopolize markets. The tangible evidence is to be found in the segmentation of the public telecommunication network.

In some quarters of the telecommunication industry, there is a perception that the technical details and choices in telecommunication network design can be resolved only by those with engineering expertise. However, the analysis of the choices involved in the design of the intelligent network in this study shows that these choices have implications for the terms and conditions of public network access. Design choices are therefore too important to be left only to the network engineers.

The implementation of proprietary interface standards in the design of the intelligent network, combined with pressures to upgrade the capacity of the public telecommunication network to meet the requirements of multinational firms, is perpetuating disparities in public network access conditions. This observation is in accord with the views of adherents of both the Strategic and the Idealist models. For example, the Department of Trade and Industry in the United Kingdom argues from the latter perspective that 'it is important that the development of network architectures and strategies for their implementation will not inhibit entry into this [intelligent network] market' (Department of Trade and Industry 1992: 2). The proponents of these models differ as to whether a technical solution – for example, a regulatory requirement to implement network interfaces – will provide an effective response. Within the Idealist model, the distortions in the marketplace are temporary. Within the Strategic model, the intensification of rivalry, for example, within the global 'Triad' – the United States, Europe and Japan – is not likely to lead to a fading away of market distortions and biases. Rather, it will recreate these distortions and biases in new forms. These dynamics of change are not amenable to a technical fix.

### *Globalization and Intelligent Networking*

The globalization phenomenon is frequently said to be driving political and economic transformations in the international economy, but it is characterized by a high degree of ambiguity. Globalization has been described as a process in which

> the internationalisation of science and technology has gone hand in hand with an increase in transnational networks and strategic alliances between enterprises as a means to competitive advantage in global markets, increasingly through the joint development of access to technology. It raises major questions about the role and grip of government policies, and their relationship to the strategies of such enterprises. (Soete 1991b)

There are those who have suggested that trends toward the globalization of markets involve increasingly footloose large corporations and a myriad of international strategies ranging from direct foreign investment to formal and informal strategic alliances. Ohmae goes so far as to argue, for example, that 'it does not matter who builds the factory or who owns the office building . . . what matters is that the global corporations . . . act as responsible citizens' (1990: 194). On the other hand, Hu suggests that the so-called global company is really a national company with some international operations and that such companies do not lose their international identity or differentiating characteristics (Hu 1992). There is on-going debate concerning the importance of national systems of innovation in the globalization process and the competitive advantages that accrue to nationally based, globally operating firms (Duysters and Hagedoorn 1992; Porter 1990; Reich 1991).

In the present context, the importance of globalization is associated with its impact on the structural and operational characteristics of firms. Many globally operating firms depend substantially on electronic modes of communication to centralize or decentralize their R&D, production, marketing and other operations. Closely associated with the globalization phenomenon is the rise of the *network corporation*. This type of corporation is based upon 'inter-company alliances of technical, production, financial and marketing competencies across national boundaries' (Soete 1991a: 52). The network corporation depends increasingly upon the public and private telecommunication infrastructure to achieve its goals. This is the case regardless of whether the goal is to strengthen competitiveness on the basis of new forms of co-operation or of oligopolistic market structures. The network corporation is characterized by formal and informal co-operative agreements among its subsidiaries, sub-contracting organizations and customers, as well as with other corporations (Duysters and Hagedoorn 1992). Soete has suggested that any of these linkages can become a bottleneck to economic growth and development.

The network corporation, with its requirements for the management and control of complex internal and external information flows, is also associated with the transformation of advanced global telecommunication networks and the growth of electronic information and communication services. The so-called network era is forecast to be in place by the late 1990s. The network era presages the widespread availability of a high-capacity, flexible, seamless telecommunication network which incorporates vast intelligence. The intelligent network is one of several design configurations that are being used to achieve this vision. This scenario is based upon an assumed rapid diffusion of an advanced telecommunication infrastructure which supports all facets of the network corporation's information-related activities. The Idealist model assumes that the benefits of developments in telecommunication become available, not only to the network corporations, but also to all their suppliers and customers as well as the individual consumer. It also assumes that this is a desirable development.

In the Idealist model, the political and economic incentives created by changes in the structure and configuration of inter- and intra-firm networks, the importance of information networks – such as electronic, non-electronic information exchange networks, and the telecommunication network – are homogenized in an unspecified process of institutional and technical change. Outcomes are the result of interacting, interdependent and balanced pressures within the three main spheres of activity as shown in Figure 10.1 – that is, the strategies of network corporations, the trajectory of innovation within the telecommunication sphere and the assumed advantages of electronic information networks. According to the premises of this model, the network corporation, its related information and communication networks, and the telecommunication network, evolve so as to complement one another. They evolve for the benefit of all other participants in the economy and society as well. When outcomes in the market do not approximate the competitive relations envisaged by the Idealist model, external explanatory factors are sought.

Figure 10.1 shows that, as conceived by the Idealist model, exogenous pressures for change are created by innovation and, regulatory and

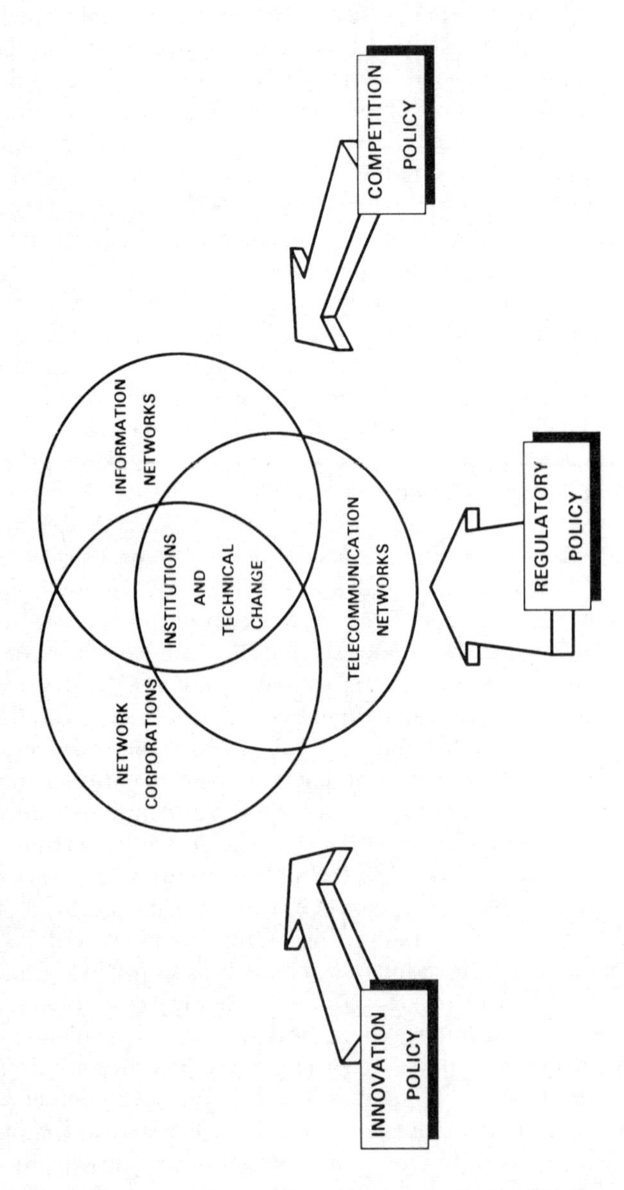


Figure 10.1 *Networks, organizations and policies*

competition policy. These are looked upon as the primary sources of distortions in the marketplace. Their eventual withdrawal is treated as a positive step towards enabling the forces of competition to eliminate the bottlenecks that create market distortions. In telecommunication, two such bottlenecks are perceived to be under-investment in advanced public telecommunication infrastructure and inadequate incentives to engage in costly research and development.

From the perspective of this model, de-regulation should be complete in order to allow the free reign of competitive forces and the subsequent removal of bottlenecks to the widespread diffusion of advanced communication networks. For example, Porter argues that an industry structure model of telecommunication in the United States which includes sellers and buyers must take into account rivalry, barriers to entry, the bargaining power of buyers, the bargaining power of suppliers and the threat of substitute products and services. He concludes that buyers in the telecommunication market are enormously powerful, very sensitive and quite sophisticated. This, he argues, is the case for *all* buyers. Therefore, regulatory interventions such as price controls are unnecessary since no firm has substantial market power. Indeed, the concept of market power is irrelevant since he argues that, within an industry structure model, the market share held by a supplier is not a good indicator of competitive advantage. The main barrier to effective competition in the telecommunication market is the time necessary for formerly monopolistic firms to learn how to compete. This is the significant market distorting factor (Porter 1992). On the basis of this assessment, he suggests that, in the US market for telecommunication, 'we have an industry structure that fundamentally supports active competition' (Porter 1992: 42). He calls for complete de-regulation of the telecommunication market.

Noam also concludes that the introduction of competition in telecommunication in the United States over the past two decades has resulted in a situation where telecommunication network provision resembles the rest of the economy. For Noam the future will be one in which:

> The network environment will be essentially a pluralistic network of user associations, a network of networks that are partly overlapping and partly specialized along various dimensions such as geography, price, size, performance, virtualness, value added, ownership status, access rights, kind of specialization, extent of internationalization, and so forth. . . . There still will be broad-based public networks . . . but just as important will be economies of group specialization, economies of clustering, and economies of trans-nationalism. (Noam 1992: 9)

In contrast to Porter, however, Noam concludes that it would be naïve to expect a reduction in regulatory tasks as a result of the changes in the characteristics of public and private telecommunication networks. In line with the Strategic model and its forecast of increasing regulation, Noam suggests that the main tasks for the future will be protection of interconnection and access, mechanisms for redistribution, prevention of oligopolistic behaviour and cyclical instability, and new institutional arrangements that match the scope of global networks (Noam 1992: 10).

The Idealist model has a high profile in much of the rhetoric concerning the emergence of a new industrial structure designated as a *network of networks*. This is forecast to have features such as those shown schematically in Figure 10.2. Telecommunication operators, providing public and privately owned network infrastructure, will create intelligent platforms which support a multitude of applications provided by competitors. Linked to the interconnected global public network will be parallel networks using radio mobile and satellite technologies. Equipment suppliers from the computing, software services and related sectors will interact with the telecommunication manufacturers to supply the components of intelligent network platforms as well as peripheral applications. They will also serve as systems integrators to alleviate problems created by technical complexity and to reduce customer uncertainty in the selection services and equipment. New entrants will continuously and effectively challenge the market power of the incumbents such as the PTOs, gradually eroding any residual ability to control access to the public telecommunication infrastructure.

On the demand side of the market, reduced barriers to entry for service providers, as well as for some network operators, will create new opportunities to differentiate competitive services for all segments of the user community. Not only will the requirements of large companies be met swiftly and flexibly, but the same advanced network infrastructure will also support applications and access for all other users. In short, the telecommunication evolutionary process is conceived as a win-win game in which all users become beneficiaries of innovations in the underlying technical system.

According to the premises of the Idealist model, responsiveness to the requirements of the network corporation and the exigencies of the globalization process requires that the design, structure and organization of the telecommunication infrastructure be revamped to reduce telecommunication costs for globally competing firms, increase flexibility in the management of information, and hasten the introduction of advanced communication services. The intelligent network architecture signifies how these goals could be met by the rapid deployment of computing and software innovations in the public telecommunication infrastructure.

Analysis of the views of network designers in the preceding chapters indicates that the implementation of the intelligent network is expected to alleviate the bottlenecks created by the historical hierarchical structure of the public telecommunication network. In the Idealist vision, it also is expected to eliminate opportunities to control the conditions of network access. The intelligent network is allied to the rise of a competitive demand-driven telecommunication industry. Since the network corporation is able to exert pressure on the PTOs and the equipment manufacturers, for some this is sufficient justification for the conclusion that all users have gained greater control over the electronic communication environment.

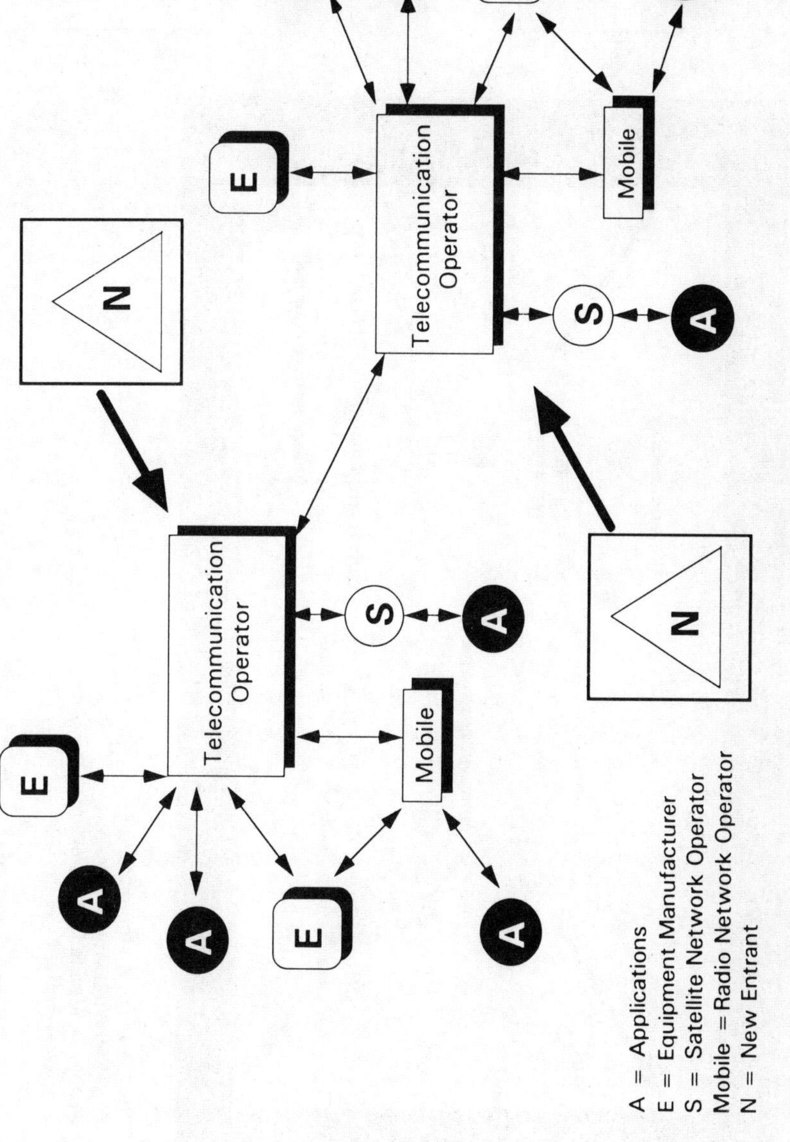


Figure 10.2 *Telecommunication transformations – Idealist model*

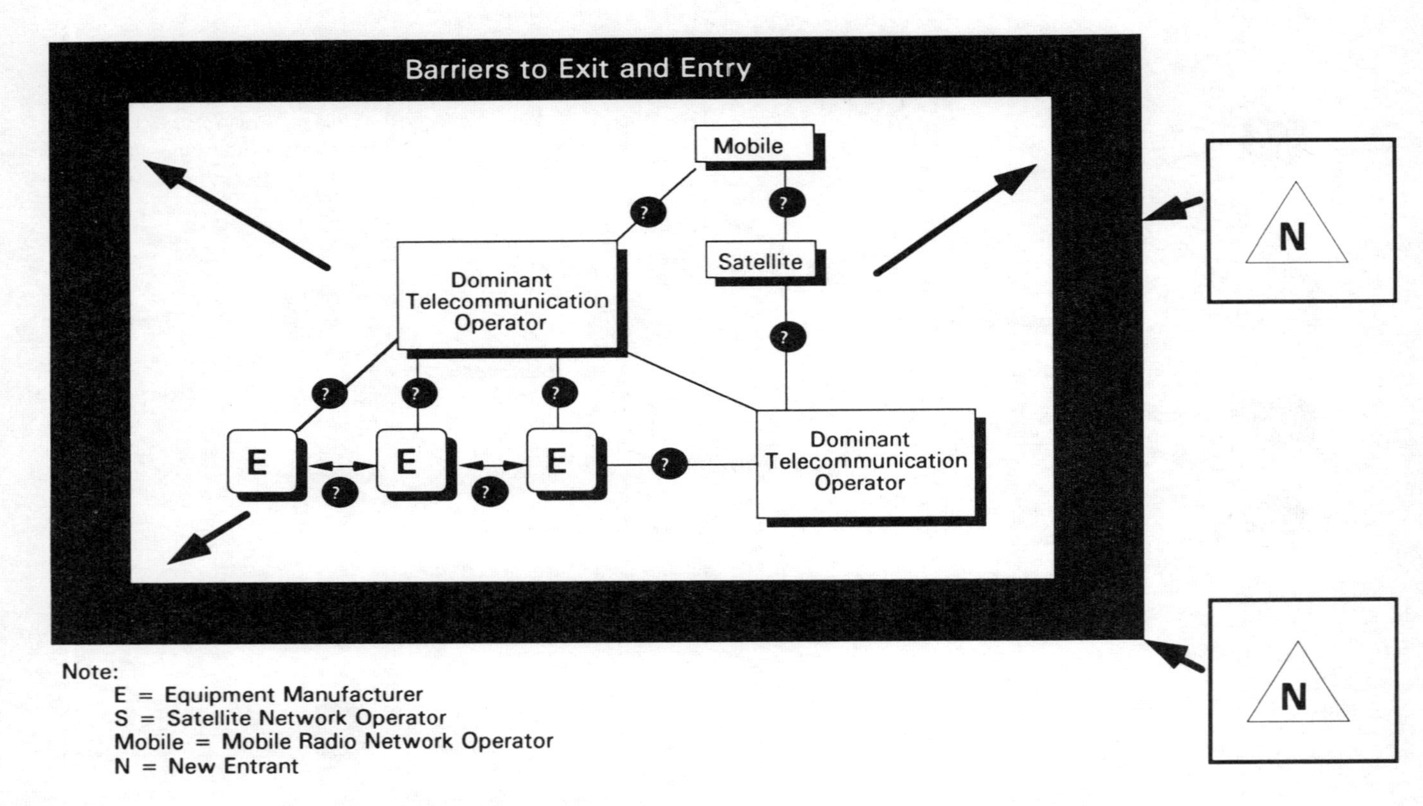


Figure 10.3 *Telecommunication transformations – Strategic model*

Figure 10.3 highlights the process of telecommunication evolution that is suggested by the Strategic model. In this model, a continuing process of monopolization and international rivalry creates new incentives for technical and institutional innovations which support the maintenance or establishment of market power for *some* suppliers and *some* users. Over time, the dominant telecommunication operator(s) may change, as suggested by the arrows in Figure 10.3. Their relationships with a set of smaller network operators who presently provide mobile or satellite networks may also change as indicated by the question marks in Figure 10.3. The relationships among the key equipment manufacturers are subject to new alliances and merger activity. In time, the major manufacturers may be represented by firms with core technical competencies in computing or in telecommunication supply.

However, from the perspective of the Strategic model, it is not only the fact that these relations are subject to change that is at issue. The extent to which firms participating in this process continue to devise innovative barriers to market entry and exit is the salient feature. This characteristic must be the basis of an assessment of the potential for the renewal of distortions in the marketplace during the process of innovation.

The design of the intelligent network in a selection of European countries has illustrated the ways in which telecommunication suppliers in Europe are influencing key aspects of the evolution of the public telecommunication infrastructure. The analysis has shown how new ways of co-ordinating the global telecommunication *network of networks* are, in fact, perpetuating the monopolization of markets and creating new barriers to network access. At the same time, the forces of oligopolistic rivalry are working to prise open access to some networks for selected users.

## The Non-Neutrality of Technical Design

The main design parameters which have guided this investigation of the evolution of the intelligent network are shown in Figure 10.4. This figure draws together the parameters that have been used to explore the implications of trends in the public network: network interface standards, unbundled intelligence, product differentiation, service competition, network access and network control. In Chapter 2, the importance of these parameters in structuring the physical aspects of the network, the management of networks and the provision of service applications was discussed in reference to transformation units that shape technical and institutional innovations. This set of parameters provides a basis upon which to assess the relative predominance of the characteristics of the Idealist and the Strategic models. Since both models are abstractions from the reality of telecommunication network development, elements of both should be present in the market. The question for policy analysis is whether the long-term outcome

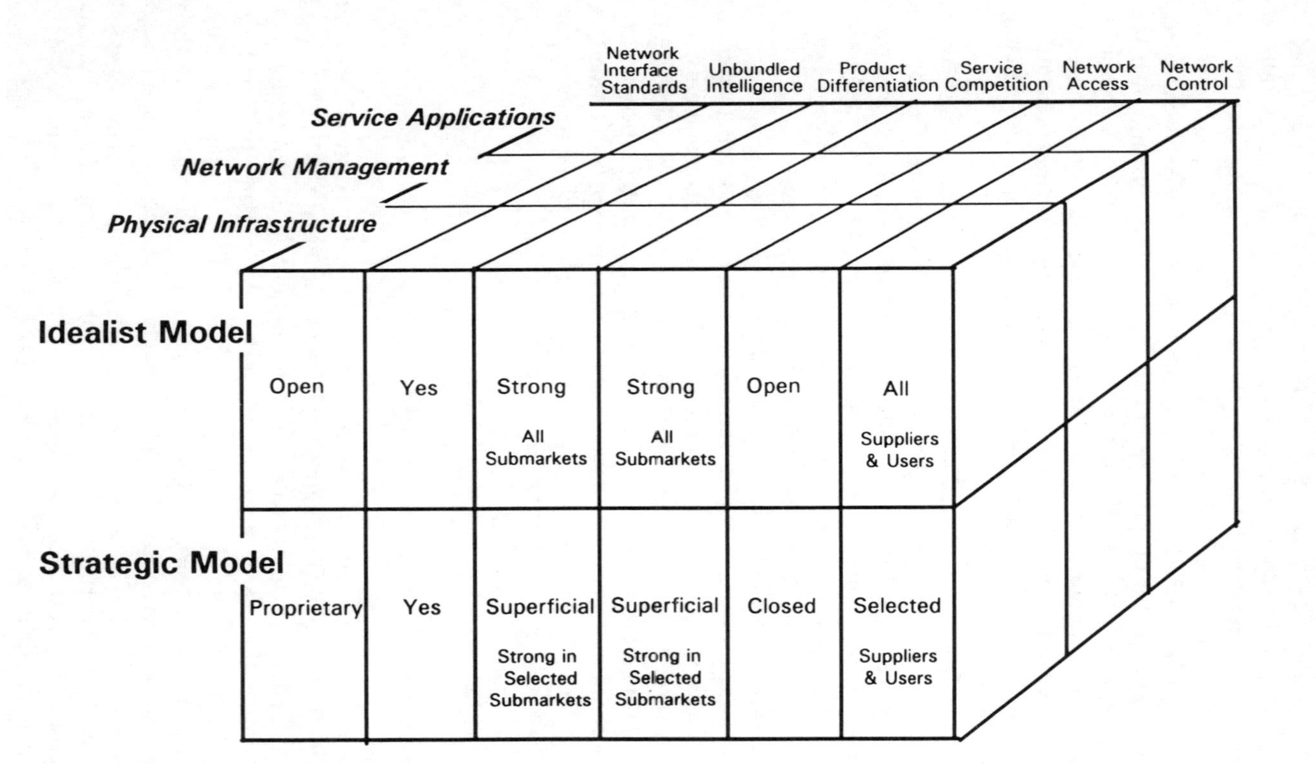


Figure 10.4 *Intelligent network design parameters*

of technical and institutional selection processes points to the continuing predominance of the Strategic model.

If the Idealist model is the closer approximation of trends accompanying network evolution, there should be evidence of the following characteristics:

*Network Interface Standards*: The trend should be towards open standards supporting interworking and interconnection of all public and private network components.

*Unbundled Intelligence*: The trend should be towards network functionality located centrally, or peripherally, in public or private networks and available from suppliers in multiple configurations at prices which reflect the costs incurred by the network operator.

*Product Differentiation*: There should be clear evidence of product differentiation based upon price, quality and service functionality rather than upon superficial differentiation in equipment design.

*Service Competition*: The trend should be towards increasing competition in all categories of service supply by customer class, application and geographical distribution and with regard to maintenance, billing, functionality and applications.

*Network Access*: The trend should be towards open, evenly geographically distributed public network access, absence of restrictions on the use of public network resources, price discrimination based on real underlying costs or quality differentiation which are subject to independent verification. Network interconnection among competing operators should be transparent and easily achieved.

*Network Control*: The trend should be towards an even distribution of network control (management, bandwidth allocation, use of network resources) between network suppliers and all categories of users. The concept of universality of voice telephone service should be in the process of being updated to take into account the potential of intelligent network resources for all users.

If the Strategic model is a more realistic approximation of the process of network evolution, the observable trends should point to an inverse set of characteristics. For example:

*Network Interface Standards*: The trend should be towards proprietary interface standards being maintained in key segments of the network infrastructure, software developed to support network management, and some aspects of service applications. These proprietary standards should be complemented by open standards where manufacturers and operators believe that such standards strengthen, or at least do not jeopardize, their competitive advantage in the market.

*Unbundled Intelligence*: There should be increasing evidence of resistance to requests from new entrants and private network operators to unbundle intelligent network resources located, centrally or peripherally, in public networks. Such unbundling as occurs as a result of public policy requirements – e.g. Open Network Provision – should tend to be responsive mainly to large user requirements. Pricing strategies should tend to be non-transparent in a bid to create advantage for the public network operator.

*Product Differentiation*: The trend should indicate superficial variations in equipment design with some evidence of strong differentiation in selected submarkets where competition is strongest, where cross-subsidies can be introduced, and/or where costs associated with innovation are highest.

*Service Competition*: The trend should indicate weak forces of competition in maintenance, billing, use of network resources and service applications in some submarkets, and strong forces of competition in others where the majority of users are large network corporations. Incentives to cross-subsidize between services where competition is strong and those where it is weak, combined with the absence of transparency in pricing, should create advantage for large users. Trends in the market should maintain or exacerbate disparities in network access for the majority of users.

*Network Access*: The trend should be towards closed and uneven geographically distributed network access, restrictions on the use of public network resources, price discrimination based on volume discounts unjustified by differences in cost or quality, and increasingly difficult negotiations over the terms and conditions of network interconnection.

*Network Control*: The trend should be towards increasing disparities in the degree of network control available to service suppliers and different types of users. The concept of universality of voice telephone service would not be updated to take into account the intelligent network resources available to users of the public network.

These alternative models of network development demonstrate that the technical aspects of the intelligent network are far from neutral. The competitive market which is embodied in the Idealist model and would erode market power is not a realistic representation of the recent experience of telecommunication evolution in the United States or in Europe. In Europe, as in the United States, rivalry is restricted to submarkets of the telecommunication industry. This applies both to equipment manufacturing and service markets. Where competitive forces do approximate the expectations of the Idealist model, this is evident mainly in submarkets where multinational firms are operating on a global or regional basis.

## The Bias of Intelligent Networking

Rivalry among the main participants in the telecommunication market is being expressed in innovative ways as the public telecommunication network evolves. Standardization and network design have become the critical transformation units which arbitrate between success and failure as incumbents and new entrants forge their strategies. Interface standards at all levels of public network operation, including infrastructure operation, network management and service applications, are the sites of strategic choices. This can be seen clearly in the example of the intelligent network. The design of interfaces permitting access to the intelligent Service Management Systems and the Service Switching Points are critical to maintaining long-term competitive advantage. The intelligent network Service Switching Points have been called the *arbitrators of control* by the manufacturers and PTOs. The Service Management Systems have been described by industry representatives as the heart or core technology in the intelligent network environment. Their design and location within the public network are considered critical to the strength of entry incentives and to the competitive prospects of manufacturers and service providers.

The design of the components of the intelligent network components is dependent upon complex software. There is evidence that proprietary software and interface standards for the implementation of the Service Management Systems and Service Switching Points are predominating over open interfaces at critical junctures in the public network. There is evidence, too, of reluctance on the part of standards-setting organizations to move to agree open interfaces when this would improve the transparency of access to public network resources. One exception is the trend towards a controlled, but expanding, degree of access to public network intelligent resources for large telecommunication customers and some private network operators. However, these developments serve to widen the gap between the degree of control available to these operators and customers as compared to that available to smaller firms and residential customers.

Closely related to the issue of standardization of network interfaces is the problem of unbundling network resources. In no case is there evidence of a willingness on the part of the traditional PTOs to provide competitors or customers with fully unbundled access to intelligent network functionality. In addition, the European PTOs show few signs of being willing to relate the costs of intelligent network functionality to the prices set for access to those network resources which they do make available. Instead, they attribute the increased costs of public network investment to the general need to upgrade the public infrastructure. The assumption is that, ultimately, intelligent services will benefit all public network users.

Furthermore, few representatives of the equipment manufacturers or the PTOs argue that a completely transparent telecommunication network infrastructure will develop so that an open multi-vendor equipment environment emerges. Failure to achieve this is attributed to the complexity of software, to the obligation to ensure network security and control, and to the need to protect knowledge and design expertise. Although all the manufacturers interviewed for this study are involved in co-operative ventures with competing service providers or with other manufacturers, critical areas of technical development are being protected by proprietary standards to safeguard competitive advantage.

The predominant trend in intelligent network design is towards increasing network segmentation. Segmentation is being redressed in selected submarkets in response to demand by large firms that are pressing equipment manufacturers and PTOs to improve the interoperability of international services by introducing gateways that bridge differences in national service implementations. This is especially visible in the flurry of activity around the establishment of international virtual private networks and the extension of service applications on a bilateral cross-border basis in Europe. However, these developments do not appear to be mitigating trends towards the implementation of proprietary standards at the national level in European public telecommunication networks.

In the past, the European PTOs could rely upon their legal or *de facto* monopoly status to forestall new entry. They established international gateways to ensure the interoperability of a limited range of end-to-end services, such as international voice telephone services, data transmission services, and, more recently, advanced services such as electronic mail, electronic data interchange, and so on. New network designs were implemented at a pace reflecting the peculiarities of their respective national markets. However, with the cumulative pressures from governments and globally operating network corporations to achieve greater efficiencies in the supply of telecommunication equipment and the public telecommunication infrastructure, the economic incentive to differentiate products and services using proprietary standards has become more acute.

The two main design parameters of the intelligent network – namely, proprietary interface standards and bundled network resources – are exclusionary tactics that are being employed to safeguard the traditional markets of the PTOs and their equipment manufacturers. The evidence in this study shows that intelligent network design choices are resulting in less open public network access than the assumptions of the Idealist model require. These design parameters have counterparts in related tactics which augment the process.[2] The most important of these tactics is pricing strategy. The oligopolistic rivalry among PTOs is resulting in a growing tendency to introduce volume discounts for large telecommunication users. These discounts coincide with the increasing costs of public network modernization. In submarkets characterized by strong rivalry the clear economic incentive is to reduce prices for network access and use. Public intelligent network resources are intended primarily to support advanced telecommunication services for these users. The evidence in France, Germany, Sweden and the United Kingdom shows that, when the PTO engineers, marketing personnel and the equipment manufacturers discuss intelligent network services, they acknowledge that most of the development costs are for services targeted to global operating businesses. These services are expected to generate substantial revenues long before intelligent network services become cost-effective for residential customers and smaller businesses (European Telecommunication Consultancy Organization 1990: 8).

With regard to network access, in most cases new market entrants must interconnect with the public network, which is controlled by a single operator, or a limited number of providers of the public infrastructure. Even in the United Kingdom where facilities-based competition in the provision of the public telecommunication network has gone farthest, the Department of Trade and Industry has expressed concern that the intelligent network will develop in a way that could prevent new operators from entering the market (Department of Trade and Industry 1992; Handford and Newman 1992).

There are very few signs that a transparent process of establishing the relevant costs and prices of network access is emerging. In addition, public policy is tending to favour telecommunication pricing strategies that embrace volume discounts and flexibility to reduce telecommunication costs for large users. The recovery of the major share of public network investment is increasingly via charges for connection and use of the domestic public network. These charges tend to be directly or indirectly levied upon all users without regard to their use of the intelligence within the network.

## Reassessing the Regulatory Goals

These trends present new challenges to European policy-makers to formulate targets for regulatory policy. The design and development of the intelligent network offers an opportunity to re-negotiate the economic and political power relations among PTOs, their equipment manufacturers and network corporations. It also creates an opportunity to re-consider the appropriate goals for regulation. As Melody has suggested,

> a major challenge for public policy will be to find methods to ensure that developments in the information and communication sector do not exacerbate class divisions in society and that its benefits are spread across all classes. This will require new conceptions and operational definitions of the 'public interest' and of public services, new interpretations of the requirements of social policy, and the design of new institutional structures for its effective implementation. (Melody 1987b: 1336)

The future structure of the global telecommunication market will be characterized by oligopolistic rivalry. The need for effective regulation will grow rather than diminish. The market power of the manufacturers, the PTOs that emerge as dominant in international, regional and national markets, and a small number of network corporations, will perpetuate distortions in the telecommunication market and sustain the relevance of the Strategic model.

The PTOs in Europe are likely to be unevenly successful in using parameters of public network design to retain control of their domestic markets, and an uneven distribution of market power in the region will make regulation increasingly complex. Just as the RBOCs in the United States are implementing different versions of the intelligent network and unbundling the intelligent network's resources in inconsistent ways, so the European PTOs will follow suit in their respective markets.

Some convergence in public network design is likely to emerge at the international level as a result of pressures from the larger telecommunication customers. However, this process is being worked out within agreements among the consortia providing virtual private network services which will support bilateral gateway agreements among the dominant PTOs and other service providers. Unlike earlier transformations in the public telecommunication network which, for example, supported gateways to enable international direct dialling targeted at the voice-telephony market, the newer developments are targeted at the heterogeneous requirements of large corporate customers. The final chapter considers whether policy-makers and regulators in Europe are well equipped to contend with the implications of the Strategic model. Will they ensure that smaller firms and residential users are not confronted with a public network incorporating intelligent features far in excess of their requirements? The answer to this question requires more sophisticated and informed policy development and regulation than was the case in the relatively simple telecommunication monopolistic environment of the past.

## Notes

1 For example, the five Nordic countries have a penetration rate of between thirty and sixty mobile phones per 1,000 population, which compares with the average penetration rate of forty-three telephone mainlines per 100 population in 1990 in the OECD area (OECD 1992: 8, 92).

2 These include, for example, costly advertising campaigns which may not be fully borne by the competitive segments of a PTO's business, reluctance to establish interconnection at points in the public network that would stimulate the growth of competition, and unwillingness to provide network design and planning specifications to potential users of the public network.

# 11
# Challenges for Policy and Regulation

> It is vital that we understand technology to be a social variable, as something that can be changed according to the choices that inform it . . . because technology is political, it must be recognized that, under current political auspices and for the foreseeable future, the new technologies will invariably constitute extensions of power and control.
>
> (Noble 1986: 351)

## Reconsidering the Policy Objectives

The Idealist and Strategic models have been considered in the preceding chapters to assess which of these models offers the best foundation for analysing the implications of the telecommunication industrial structure in Europe. The Idealist model has been found wanting. It is visible in the rhetoric of some representatives of the policy, telecommunication supplier and large user communities. Representatives of these groups often have political or economic interests in championing the design and implementation of an open public intelligent network. However, they also have strong countervailing interests in public network closure. The Strategic model embraces the reality of the contradictions underlying technical and institutional change in the telecommunication industry in both the United States and Europe to a greater degree than the Idealist model. The former model points towards the continuing uneven evolution of the public network and its associated institutions.

There is, therefore, a need for effective policy intervention to counter the disparities in network access conditions that will continue to arise. This requires that the power of the state be exercised through the public regulatory process. This chapter considers whether European institutions are likely to engender an environment that is capable of ensuring that the public interest in telecommunication embraces the concerns of smaller firms and residential consumers in the face of strong pressures to redefine that interest more narrowly in terms of the strategic priorities of a relatively small number of multinational suppliers and users.

The focus in this chapter is mainly on the political and economic factors contributing to the design of emerging policy and regulatory institutions. The detailed procedures and methods of regulation are part of the continuous struggle for control which is mirrored in the formation and dissolution of public and private institutions, and in the appropriation, or rejection, of technical designs. It is assumed that policy and regulatory institutions can be designed so as to maximize the equitable distribution of the socio-economic benefits of innovations in the technical artefacts of communication (see Smythe 1979). The design of the intelligent network is one instance in this struggle for control which is shaping the trajectory of public network development.

### *The Public Interest in Telecommunication in the United States*

As a result of the constellation of global, regional and national rivalries among firms which both produce and use the intelligent network, advanced electronic communication systems are being designed in the light of goals that resemble those in much earlier phases of telecommunication development. These are, on the one hand, to promote the apparent coincidence of the development of integrated, increasingly advanced telecommunication systems with the public interest in universal services; and on the other, to facilitate the segmentation of networks to meet selected business requirements.

For example, from 1876 to 1893, in the earliest phase of telephony in the United States, a regime of monopoly supply was predominant. With the expiration of the Bell patents in 1893–94, competition existed until 1920 when AT&T achieved a *single system* as a result of a programme of acquisitions and monopolization. Private monopoly then became the dominant form of industry structure in the United States. During the early competitive period, services were extended rapidly and there were substantial reductions in tariffs for business and residential service users. Nevertheless, in the competitive period much of the development of the public network occurred in the business core areas of large urban communities (AT&T 1910: 23–4). Later, though independent telephone companies were numerous in rural areas, the emphasis under monopoly shifted to a convergence of the monopolistic industry structure with the *public interest* in the development of a ubiquitous public telecommunication network. Concepts of system integrity, unitary planning and economies of scale were espoused by the privately owned monopolist, AT&T. AT&T sought to achieve universal service on the basis of one system and one policy throughout the United States (Gabel 1969).

Public utility regulation for telecommunication, first by the Interstate Commerce Commission (1910–34), and then by the Federal Communications Commission, emerged partly in response to the AT&T President's contention that 'if there is to be no competition, there should be public control' (AT&T 1907: 18). Regulatory institutions were also established in the belief that public regulation of private monopoly could ensure the allocation of economic resources to developments in telecommunication which would reflect all user requirements and avoid the excesses of monopoly power.

The public telecommunication operators were considered to be *businesses affected with the public interest* and their activities of public concern. This principle has underpinned the approach to the regulation of public utilities in the United States, and its interpretation has provided the justification for public intervention in the business affairs of private companies such as AT&T. This concept of the public interest was rooted in British common law. The principle was established by Lord Chief Justice Hale in his treatise *De Portibus Maris*, such that when private property is 'affected with a public interest it ceases to be *juris privati* only'. This statement was reaffirmed by the US Supreme Court in a decision in 1876. In Mumm v. Illinois, Chief Justice Waite argued on the basis of Lord Chief Justice Hale, some 200 years earlier, that

> Property does become clothed with a public interest when used in a manner to make it of public consequence, and affects the community at large. When, therefore, one devotes his property to a use in which the public has an interest, he, in effect, grants to the public an interest in that use, and *must submit to be controlled* by the public for the common good, to the extent of the interest he has thus created. He may withdraw his grant by discontinuing the use; but so long as he maintains the use, he must submit to the control. (United States 1976; emphasis added)

The complex system of Federal and state regulation of telecommunication services which evolved in the United States was honed to the changing circumstances of network development in that country and to the particular relations between the state and the private sector. This structure of regulation produced detailed procedures for administrative oversight, such as methods for calculating revenue requirements to meet AT&T's allowable rate of return, cost accounting methodologies for assessing whether tariffs were fair, reasonable and non-discriminatory, and for determining the magnitude and direction of internal cross-subsidies (Melody 1987b). It built a vast edifice of information requirements and a bureaucracy of legal experts, economists and public interest advocates. However, the public switched telecommunication network did not keep pace with large telecommunication user requirements, nor did it adequately serve the needs of rural and remote users in the United States. Larger users were often served by private line networks using facilities provided by AT&T and, later, by new entrants when the forces of technical, regulatory and market restructuring took hold. Customers in rural areas, in many cases, were served by publicly subsidized, independent rural telephone administrations.

At the Federal level in the United States, the regulatory regime produced a concept of universal service which the FCC described as follows:

> Access to telephone service has become crucial to full participation in our society and economy which are increasingly dependent upon the rapid exchange of information. In many cases, particularly for the elderly, poor or disabled, the telephone is truly a lifeline to the outside world. . . . Our responsibilities under the Communications Act require us to take steps, consistent with our authority under the Act and the other Commission goals. . ., to prevent degradation of

> *universal service* and the division of our society into information *haves* and *have nots*. (Federal Communications Commission 1984: para 9; emphasis added)

This statement of concern for the information 'haves' and 'have nots' has become synonymous with the public interest in the American telecommunication industry.

From the earliest days of telephony to the current generation of public networks, the perception of correspondence between the dominant telecommunication suppliers' interpretation of the public interest and that of the state has varied with the priority given to national security, implicit and explicit industrial policies, and new definitions of universal service. Today, as throughout the history of telecommunication supply (Bonbright 1962; Posner 1969; Trebing 1969b), debate continues over the reconciliation of different perceptions of the public interest in telecommunication supply on the basis of reasonably independent standards of equity and efficiency.

### *The Public Interest in European Telecommunication*

In Europe, government postal administrations gained control of telegraphy by the 1860s, and, in most countries, the telecommunication industry evolved as a system of nationally, regionally, or municipally based public monopolies sanctioned either by law or by the common consent of the state. Although the regulatory treatment of telecommunication in the United States had been derived from British common law, the telecommunication operators in Europe were predominantly state-owned and there was no need to develop a separate regulatory apparatus to contend with private businesses that might be *affected with the public interest*. The concept of universal service which had evolved in the United States had its counterpart in Europe in a rather different concept of *public service*. As Garnham has observed:

> The concept of public service is stronger in those [European] countries governed by Roman law rather than common law and derives from a Hegelian tradition which distinguishes the State, as the expression of the highest form of social rationality and of the public interest, from civil society, the subordinate realm of competitive private interest. Within this tradition the State, by definition, represents, through the political process, the best interests of all citizens. Thus the delivery of a public service by the State, whether directly or by delegated authority, does not require a more specific *universal service* remit nor is there a requirement for the State to be held accountable for its actions, legally, or otherwise, to individual citizens. (Garnham and Mansell 1991: 29)

Garnham has argued that in the European Commission's 1987 Green Paper on Telecommunications, two concepts of universal service were present (Commission of the European Communities 1987). The first is a concept of universal geographical availability.

> This concept is closely related to the development of nation states as political and economic realms with distinct territorial identities . . . all citizens, wherever they

> are located within the boundaries of the State, have a right to equal access to that State's services. At the same time the provision of such services by a central power, especially infrastructural and communication services, binds the periphery to the centre with the aim of national political and economic unification. This is a concept of pure access. (Garnham and Mansell 1991: 26)

The second concept is related to non-discriminatory access or 'the equal treatment of all users in terms of price and/or levels of service' (Garnham and Mansell 1991: 26). As a result of these differences in the articulation of the public interest through publicly owned or franchised PTOs, no tradition of independent regulation emerged in Europe, at least not until pressures for liberalization of the sector began to be felt.

A new regulatory institution was created with the introduction of competition in the United Kingdom in 1984. The European Commission's agenda of liberalizing the telecommunication market led to its announcement of an intention to seek the independence of regulatory functions from the operations of PTOs in 1987 (Commission of the European Communities 1987). The European Community member states began to introduce procedures for the regulation of European PTOs, but the powers, procedures and the degree of separation of regulatory institutions from the operations of the PTOs have differed with the political and economic environment in each of the member states (Mansell 1989).

This relatively late arrival of debate as to the form and substance of regulation in the telecommunication sector in Europe as compared to the United States brought some clarification of the goals of telecommunication policy. On the one hand, within the framework of European Community legislation, policies and regulations were to be devised by the member states to promote the efficiency and competitiveness of telecommunication supply. On the other, universal telephone service was to be promoted through the harmonization of standards, the promotion of geographical availability and non-discriminatory access to public networks. However, according to Garnham, the concept of universal penetration of telecommunication services on the basis of reasonable costs and affordability remained absent from Europe in public debate and in the formal texts of European policy and regulation (Garnham and Mansell 1991).

In the United States, a highly specific regulatory apparatus emerged which has found it extremely difficult to adjust to the complexities of the domestic and international telecommunication environment. In Europe, the institutions of telecommunication regulation are still in an embryonic state. The European Commission has exercised its powers under the Treaty of Rome as well as its powers of indirect political and economic persuasion to introduce a more liberalized telecommunication market and to promote the emergence of independent regulatory institutions. We look first at the rhetoric which has accompanied changes in European telecommunication and then at the regulatory apparatus that is being deployed in an attempt to implement policy objectives.

*European Telecommunication Policy Rhetoric*

The European Commission announced in 1983 that 'telecommunications today have an economic, social, cultural and natural strategic importance which will mark them out for a large-scale initiative' (Commission of the European Communities 1983: 7). The Commission went further, to state that

> the recourse to the European dimension, which is recognized as a condition of success, implies a gradual transfer of powers and means to the Community. Such a transfer in no way implies any modification to the statutes or responsibilities of the national PTTs. Neither does it affect the way in which each member state organizes the transfer of financial resources between PTT and governmental budgets [footnote omitted]. . . . A Community political and legal framework, which is clear and precise, thus becomes indispensable. . . . A legal framework does not imply, however, additional constraints and bureaucracy; on the contrary, it will quickly become apparent that the gradual transfer of power and resources to the Community, if brought about as the Commission envisages, will be counterbalanced by a *reduction in regulations* and, moreover, a more rational utilization of the public resources allocated to this sector. (Commission of the European Communities 1983: 9–10; emphasis added)

In spite of its commitment to a reduction in regulation, the Commission has introduced a considerable number of directives, decisions, regulations, recommendations and resolutions in pursuit of its goals. From 1984 to mid-1992, thirty-five of these were presented dealing with liberalization measures affecting terminal equipment markets, procurement practices, service competition, open network access and mutual recognition of licences.[1] Pending the outcome of a review of the liberalization process over almost a decade which began in 1992, the Commission's view was that

> Telecommunications are an essential feature and instrument of modern society. At work and in the home the widespread availability of modern, efficient and cost-effective telecommunications contributes to competitiveness, to reducing the impact of distance and opens up new opportunities for personal and business development. . . . The 1990 directives [on Services and ONP] reflected the major political compromise achieved in the Council . . ., which was possible through *a careful balance of harmonisation and liberalisation*. The Commission remains faithful to that spirit. . . . Great care has been taken to ensure that the options explored do not call into question the need to *maintain, and perhaps even extend, universal access to the voice telephone service and network*. (Commission of the European Communities 1992a: 3; emphasis added)
>
> While there are substantial differences between Member States in terms of intensity of usage, technology and inward service, there are powerful forces driving telecommunications development everywhere in a common direction. In the context of the internal market there is a need for both *harmonisation and liberalisation* at Community level in order to ensure that the development of telecommunications across the Community is not impeded by national barriers or practices which are incompatible with the achievement of the objectives of the Treaty [of Rome]. (Commission of the European Communities 1992a: 7; emphasis in original)

Alongside its measures to stimulate harmonization and liberalization, the

Commission has argued that 'maintaining and expanding universal *access* to the voice telephone service and network must be seen as a major policy goal for the sector' (Commission of the European Communities 1992a: 31). Furthermore, as liberalization of the European market progresses, there is to be more freedom for co-operation among European PTOs within the guidelines established by competition policy for the sector (Commission of the European Communities 1991b). The Commission's programme of reform is expected to bring the full liberalization of private networks for services including voice, data and text. It appeared in May 1993 that the full liberalization of the telecommunication supply industry, including voice telephony and the public network infrastructure, might be on the verge of gaining a consensus throughout the European Community. The Commission was arguing for the full opening of the market by 1998, although this target was still being negotiated.

The texts issued by the European Commission concerning telecommunication liberalization refer to extra-Community threats created by international rivalry in service and equipment markets (Commission of the European Communities 1991a, 1992a, 1992b). When it looks to external markets, the Commission and policy-makers within the member states seem to recognize the reality of the Strategic model. They recognize that full reciprocity in the liberalization of markets within the Triad of the European Community, the United States and Japan, is far from a reality and that there are multiple indications of market-distorting practices. However, within the context of intra-Community policy, the Commission's rhetoric and actions often suggest that the Idealist model is rather more influential. Within the Community, the conflicts between harmonization and competition, and between success in global markets and the promotion of universal services, are rarely problematized explicitly.

There are strong trends in the evolution of intelligent networks towards the segmentation of networks. The PTOs, some equipment manufacturers and even some users have an interest in maintaining disparities in the access conditions for public networks. Policy-makers are espousing the benefits of open public networks and initiatives to ensure the universality of voice telephone services throughout the European Community. At the same time, they are encouraging equipment manufacturers and telecommunication operators to design and implement leading-edge public network services that exacerbate network segmentation via the use of proprietary standards. In the face of these conflicts in the rhetoric of policy and in the practices of suppliers, there is a need for more effective policy and regulatory institutions.

The European Commission publicly suggests that the tendency toward public network closure can be counterbalanced by negotiating among the interests of the large network corporations, smaller firms and residential customers. To achieve open network access, the PTOs and manufacturers must relinquish their control over the public network. The Commission has looked to the specification of a set of minimum technical and administrative conditions of network supply as one of the main mechanisms to achieve this goal (Konig 1989: 5).

Among these conditions are transparent and open access to the public network infrastructure regardless of whether the network is provided under monopoly or competitive conditions. Network interface standards and pricing policy have been adopted as the main tools of regulation. These, together with detailed accounting procedures, are expected to provide the benchmarks against which the public interest in public network development can be assessed. These tools of regulation are expected to provide a way of coping with the unequal distribution of market power within the telecommunication marketplace. Even in member states where the PTO retains special or exclusive rights to provide the public telecommunication infrastructure, all those who wish to access and use the network are to receive transparent equitable treatment. For instance, member states that maintain special or exclusive rights are expected to use the necessary measures to make the conditions governing access to the network and its functions, public, objective and non-discriminatory (Konig 1989: 6).

Representatives of the Commission have insisted upon the formulation of harmonized standards for essential public network requirements and have emphasized areas where communication across different networks requires agreement upon interface standards. For example, the international harmonization of standards is treated as a prerequisite for optimal competitive conditions at a world-wide level that will be to the advantage of consumers, service providers and manufacturers (Berben 1991: 23). Publicly agreed standards are intended to ensure the integrity of the public telecommunication infrastructure, facilitate competition and promote the interoperability of services. Furthermore, rivalry in national, regional and global markets is not to be permitted to 'jeopardise the availability of a universal set of services' (Berben 1991: 24), and the current public telecommunication network and telephone service are prime examples.

This is the rhetoric of evolving telecommunication policy and regulation in the European Community context. It carries with it assumptions that are reminiscent of both the Idealist and the Strategic models. For instance, on the one hand, the harmonization of a minimum set of standards and regulations is expected to secure the widest public interest in telecommunication development. On the other, there is an awareness of the disadvantageous impact of rivalry and monopolization and the need to secure the provision of a set of universal services.

### *The Regulatory Tools – Standards and Pricing Policy*

The tension between the Idealist and the Strategic models in the design of policy and regulatory institutions in Europe is particularly visible in the European Community's directive for promoting fair and equitable access to public telecommunication networks; that is, the Open Network Provision (ONP) Directive (Council of the European Communities 1990b). A report on progress toward the implementation of this Directive quoted a long-time observer of the liberalization process as follows:

> A European regulatory authority for telecommunication is way ahead of reality. Even if it did happen, it would have to rely on local and national authorities to do most of the work. Put another way, unless there are proper independent national regulatory authorities, there will be no enforcement and implementation. (Analysys 1991: 106)

The powers of implementation and enforcement of the European regulatory apparatus have been linked to the viability of ONP as the means of balancing conflicting political and economic interests in the evolution of European telecommunication markets. Proposals for the mutual recognition of operating licences awarded by the member states of the European Community,[2] and the establishment of a Community Telecommunication Committee (CTC) are also being considered as methods of ensuring that the rhetoric of telecommunications policy is turned into practice (Commission of the European Communities 1992c). However, the Committee would be composed of national authorities and would act within boundaries set by the interpretation of the *subsidiarity* principle (Gilhooly 1991a).[3] There are few signs that a European regulatory institution that is independent of the parochial interests of the member states or their national PTOs will emerge within the present structure of the telecommunication market in Europe.

The application of ONP to the lines leased from the PTOs to provide the underlying capacity for many corporate networks and value-added services was agreed in June 1992. It illustrates the conditions that the Commission is seeking to enforce in order to encourage equitable terms of access to the public telecommunication network (Council of the European Communities 1992b). For example, restrictions on access to, and use of, leased lines must be objectively justified; a harmonized set of leased lines must be offered with defined network termination points; tariffs for leased lines must be based on objective criteria and follow the principle of cost-orientation. Tariffs must be transparent, published, and sufficiently unbundled to meet the European Community's competition rules. They must be non-discriminatory and guarantee equality of treatment of all network users. Charges for access to, and use of, leased lines must take account of a principle of fair sharing of the cost of the resources that are used and the need for a reasonable level of return on investment. In addition, telecommunication network operators must implement transparent cost accounting systems which can be verified by accounting experts.[4]

The harmonization of the technical conditions for advanced telecommunication networks is linked with the European Commission's attempt to ensure that the increasingly oligopolistic telecommunication sector does not weaken the immature forces of competition. As the then head of the telecommunication section of the Commission's Directorate General for Competition Policy put it in 1991, the PTOs' role consists not only of the provision of a domestic network with universal coverage but also of the realization of a pan-European network through harmonized national public networks. This requires harmonization rules at a European level, such as ONP, and also co-operation agreements between PTOs (Ravaioli 1991: 62).

These developments do not reveal the underlying issue of whether the political or economic power exists to resolve conflicts among policy goals or to enforce regulations. As we have seen in preceding chapters, market distortions are not being competed away, they are giving rise to new distortions such as those visible in the design of the intelligent network. The European Commission is putting regulatory procedures into place that replicate those which have emerged in the United States. For example, in addition to the use of Directives to encourage open access to the public network, there are Directives which call for cost-orientated pricing and accounting methodologies that require judgement in the allocation of costs, and which depend upon information provided by the organizations that are to be regulated.[5]

If the Commission only pursues these modes of regulation it will recreate the features of traditional public utility regulation which have been tried and found wanting in the United States. The present trajectory of regulation in Europe assumes that open access to the public network can be achieved via negotiations which aim to encourage suppliers to be more efficient and equitable in their treatment of *all* customers. This approach will lead to institutional innovation only to the extent that it translates American modes of public utility regulation into the European context.

The telecommunication equipment suppliers and the PTOs are not waiting for regulatory decisions before they proceed with the modernization of the public network. They are investing in proprietary technical systems such as those which comprise the intelligent network. Pricing strategies are being introduced which permit ever-increasing flexibility. Consensus upon what constitutes a cost-related price continues to be elusive within the member states as well as at the Community level. Experience in the United States, and increasingly in the United Kingdom, where European market liberalization has been most extensive, has shown that the specification of costing methodologies by public authorities provides no guarantee of consensus on the interpretation of appropriate data inputs or on the resulting cost analyses. The principles of average pricing and non-discrimination that coincided with public service goals in the past are being challenged, and they can be expected to become issues in all the member states as liberalization progresses.

The Commission has no powers to engage in formal regulation of public utilities on a Community-wide basis. For example, although there are European standards in name, there are none in practice, since those which have been agreed must be implemented through national organizations (Hawkins 1992a). The European Commission has barely started down this regulatory path. The future may see the proliferation of multiple European regulatory regimes adopting practices that have increasingly little impact on the strategies of the PTOs and their equipment manufacturers.

**Minimum Conditions for European Regulation – the Policy Issues**

The fact that PTOs are developing innovative ways of retaining control over the public telecommunication network is not disadvantageous for all users. The complexity of public and private telecommunication networks and the diverse requirements of users call for a re-assessment of the role of policy institutions. This should be rooted in the reality of European telecommunication markets which is characterized by the Strategic model. If it is based upon the premises of the Idealist model, it will produce policies and regulations that are unable to redress the distortions in the market and disparities in the terms and conditions of access to the public network. In fact, it may even exacerbate them.

From the Idealist model's perspective, the European Commission's challenge is to walk a tightrope between the merits of competition and co-operation in the development of European public and private networks. However, in reality, the Commission faces the challenges presented by the Strategic model. It must guide the design of regulatory institutions and processes to contend with oligopolistic rivalry and its consequences for telecommunication suppliers and users. In this context, the issues are the following:

1. For whom should the advanced public telecommunication network be designed?
2. Are incumbent firms likely to give way to new industrial structures for the supply of public and private telecommunication services that preserve a broad definition of the public interest in access to the public network?
3. Do opportunities exist to create policy and regulatory institutions with the political legitimacy and economic power to control, or at least influence, the activities of globally operating telecommunication network suppliers?

*Priorities for Public Network Design*

The first issue concerns the priorities for public network design. On the basis of the insights developed using the Strategic model, it can be concluded that the public network should be designed primarily to meet the requirements of *all* its potential customers. It must offer a common public infrastructure that incorporates a set of minimum technical conditions. The European Commission is pursuing this goal, but it is doing so under the shadow of the Idealist model of telecommunication network development.

The Commission's Directives and related activities are intended to ensure the provision of universal service, which is defined as consisting of the 'provision and exploitation of a universal network, i.e. one having general geographic coverage, and being provided to any user or service provider upon request within a reasonable period of time at affordable prices'

(Commission of the European Communities 1992a: 23). Community-wide investment to achieve this goal by the year 2000 has been estimated at some 400 billion ECU for capital investment.[6] To upgrade the public infrastructure in the Less Favoured Regions of the Community to the standard available in the rest of the Single Market, it has been estimated that some 40 billion ECU will be needed by the year 2000.[7]

These estimates give an indication of the scale of expenditure that will be necessary if participants in the electronic communication environment are to have reasonable access to the public telecommunication network. However, as Garnham has suggested, the concept of service universality applies mainly to geographical coverage and non-discriminatory access in Europe. It does not consider the reasonableness of the costs of access or the affordability of access (Garnham and Mansell 1991). Furthermore, it leaves the resolution of this issue to highly politically charged interpretations of cost-orientated pricing relationships and the specification of technical standards for the components of the public telecommunication network.

This approach is likely to result in the escalation of costs of public network access. The trends in public network design are towards the location of costly intelligent components within the public network in an effort by the PTOs to meet the requirements of a relatively small number of large telecommunication customers. The public network is not being designed in the light of the minimum technical conditions for a *universal* network which are found in the rhetoric of policy-makers and the suppliers and large users. Historically, leading-edge users in the United States have channelled their resources to provide private network solutions to their telecommunication needs when the public network operators have been unable to do so. In most of the European Community member states, these users have been prevented, or discouraged, from opting for this strategy, not because of a failure of technical innovation, but because of the way the *single system* solution for telecommunication supply has been conceived and continues to be protected. In the liberalized marketplace of the Idealist model, the large users would opt for public network supply in preference to private supply to the degree that it offers a good fit with their needs *and* when it is accessible at a lower cost than self or third-party supply.

The European PTOs have a strong incentive to provide the advanced services for these users using an integrated public network because they can rely on smaller businesses and residential customers to contribute disproportionately to the costs of network modernization. Their incentive to pursue this option also depends on the ability to control network access by implementing proprietary interface standards and by manipulating prices of network access such that the cost of using the public network is greater for competitors than it is for themselves. This is the lesson of the Strategic model.

In the present market environment, the Strategic model suggests that large firms will shape the development of the public network by creating pressures for the implementation of advanced intelligent networks at the lowest possible cost to themselves. This is not a conspiracy or the result of collusion among PTOs, manufacturers and network corporations. It is a strategic response to the internationalization of markets and to the economic and political factors that are shaping the technical design of public networks. The telecommunication requirements of the large users will outstrip the average quality and capability of the public network as they have historically. Discrepancies between the minimum conditions – that is, standards and pricing structures – that will promote the continuing evolution of a universal public network catering to the majority of users, and those needed to meet the requirements of the network corporations, will continue to grow. There is clearly a need for radical changes in policy and regulation in the face of this trend.

The idea that technical progress and innovation in telecommunication networks are uniformly beneficial to all facets of society and the economy must be challenged. Noble has argued that 'the new technologies will invariably constitute extensions of power and control . . . they not only must be viewed with scepticism and suspicion, but perhaps must also be resisted and rejected' (1986: 351). A realistic assessment of the importance of electronic communication in everyday life suggests that not all the advanced technical artefacts of communication should be rejected. What needs to be abandoned is the Idealist model's rhetoric which camouflages the way in which public networks are designed and implemented.

In fact, two mythical notions must be rejected. The first is that the rapid diffusion of the most advanced public network capabilities is in the interests of all telecommunication users. This is the 'trickle-down' rhetoric that characterizes the dominant interpretation of the processes of development in the communication sector more generally (Mansell 1982). The second myth that should be abandoned is that technical trajectories are narrowly circumscribed by options such as those debated by the technical engineers charged with designing the intelligent network. When the designers of telecommunication networks speak of network integration or segmentation, their debates are framed by a relatively narrow range of design alternatives that are responsive mainly to the requirements of the most sophisticated users. At present, the forces of technical and institutional change are resulting in the segmentation of the public and private telecommunication infrastructure, and the dominant trend in public network design is to embed costly intelligence in the core of the public network. This is the outcome of the political economy of oligopolistic rivalry in global telecommunication markets.

However, as this study has shown, the trajectory of network development is shaped in ways that are not visible when a narrow, and relatively rigid, technical selection process is assumed. This trajectory is malleable and it can be shaped by social, economic and political action. The public network, for instance, could be designed to fulfil the most advanced requirements of the large users *or* to meet a set of minimum conditions which reflect different criteria that are in line with a broader definition of the public interest.

*Forces of Institutional Change*

The first policy issue concerned the priorities for public network design. The second issue is whether the PTOs and the telecommunication equipment manufacturers will give way to a new industrial structure which preserves a broad definition of the public interest in access to the public network. The evidence in this book suggests that if the public network is not to evolve mainly as a reflection of oligopolistic supplier power and the bargaining power of the network corporations, public policy intervention will continue to be required. However, the use of open network interface standards and the regulation of prices as the main tools of public policy will be an insufficient response to the unfolding distortions in the market. These will need to be coupled with structural change and with a transparent assessment of the ways in which new generations of technical designs – for example, the intelligent network or radio-based services – can be used to respond to the wider public interest in public network access.[8]

The PTOs and the telecommunication equipment manufacturers are shaping and appropriating the public network in the 1990s to a degree that has not been witnessed in the history of telecommunication development. The political economy perspective which has been developed in this study has demonstrated, however, that this process is not a technical juggernaut. Policy-makers can seek opportunities within the interaction of the multiple forces of technical and institutional change to ensure that the public interest in public network development is considered, if not protected, more effectively than it has been so far.

Telecommunication policy and regulation in Europe are addressing a number of tasks to alter economic incentives in telecommunication markets. Liberalization measures are being introduced, as we have seen in earlier chapters. Competition is being encouraged, but so too are co-operative activities. The goal is, on the one hand, to build an advanced pan-European infrastructure that meets the needs of multinational users and, on the other, to upgrade the public network to the average quality across the Community. As a result, despite the emphasis on competition and unfettered markets for telecommunication supply, a large proportion of public policy is aimed at controlling, or at least influencing, the activities of PTOs, and indirectly, the equipment manufacturers, whose strategic visions are focused upon the international market. As the Chairman of British Telecom has argued, ‘an increasingly interventionist regulatory approach is doomed to failure, if only because the complexity and sophistication of the market will be such as to put it beyond the wit of the regulators to regulate’ (Vallance 1993). The existing national and regional regulatory regimes in Europe are being rendered impotent in the face of global rivalries and complex networks which span the world.

The policy objective, therefore, must be to establish regulation where the impact of public intervention can be most effective. Instead of seeking to control the activities of oligopolistic firms, their international and domestic long-distance operations should be structurally separated from the domestic public access network. By unleashing the international suppliers of advanced services to confront the full force of oligopolistic rivalry in international markets, these firms would have the freedom and flexibility they seek. They would also be unencumbered by all but the most minimal restrictions on their business practices. They would no longer control the domestic public access network that is both costly and necessary to permit access to traditional and advanced telecommunication services. There are alternative modes of structural separation that could be considered, and these would depend upon the particular market conditions prevailing in each of the member states. However, this step would still not provide a sufficient response to the forces of change in global telecommunication markets.

The criteria for the technical design of a set of advanced universal services would also need to be established through a consensus facilitated by national and European-wide regulatory authorities. Regulators would actively address the problem of what features of, for example, ISDN or personal mobile services need to be accessible to smaller firms and to residential consumers. An appropriate set of specifications would need to be established, revised periodically, and enforced with the goal of optimizing access to a clearly established set of public network services. With the international service suppliers free to pursue the global market on the basis of their respective competitive strengths, the costs of network modernization would become more transparent than they can be in vertically integrated domestic markets. On the basis of this information, decisions could be made with regard to the desirability of public investment in national or regional infrastructure projects and service applications. Attention could also be given to the need to create incentives for increased private investment in the public access network in line with developments in other regions of the Triad.

### *The Creation of New Institutions*

The third policy issue is whether it is feasible to consider the creation of new policy and regulatory institutions in Europe which might have the political legitimacy and economic power to control the activities of globally operating PTOs and their suppliers. Given their strong international orientation, it will not be long before some European PTOs are reluctant to bear any public service responsibilities. However, they will argue a strong case for the maintenance of these responsibilities as long as they preserve control over the terms and conditions of public network access. This apparent synergy between a broadly defined public interest in public networks and the interests of the globally operating PTOs, the equipment manufacturers and the largest users, resembles that which has existed historically within domestic markets. In the past, the public and private telecommunication nopolies argued the centrality of electronic communication in cultural, social, political and economic life. The costs of their network modernization activities were driven by the most advanced services, and they were generally disproportionately recovered through tariffs charged to those who required access to the public network, if only to make a local telephone call. Some form of public control of the market power of the monopoly PTOs has always been deemed necessary either via state ownership, franchising or public regulation of private *businesses affected with the public interest*. In the global oligopolistic market environment of the 1990s, such public control is essential if the gap between those included, and those excluded, from the electronic communication environment is not to continue to grow.

As telecommunication liberalization takes hold, the more ambitious PTOs and the equipment manufacturers are pursuing their goals in international markets. These activities can only indirectly be influenced by nationally or regionally based government authorities. However, this does not mean that opportunities at the national or regional level to implement effective policies and regulations have vanished. The changes in telecommunication markets of the late twentieth century signal the need for the policy and regulatory regimes in Europe to initiate more than the infrastructure investment and pre-competitive research and development programmes, and Directives which have been the focus of public policy thus far. It will call for a redirection of public resources and political momentum to create new institutional regimes for European-wide regulation. A European regulatory institution is needed to create incentives for public network evolution that are aimed at reducing disparities in the terms and conditions of public network access. If steps are not taken to establish more effective regulatory institutions, the public network will increasingly benefit a small club of globally operating, oligopolistic firms. Their conception of the public interest in telecommunication will be more narrowly defined than that of the monopoly PTOs of the past.

The goal of a European regulatory institution would be to promote the ubiquity of the voice telephony network and to establish a consensus upon the intelligent network and related service features that must be promoted to meet the needs of smaller business users, public-sector services and residential consumers. Braudel observed that the pace and ease of movement within networks was conditioned, historically, by the minimum characteristics of the elements present across the transport networks that supported trade in goods (Braudel 1984: 25). There is an analogy with the public telecommunication network of the 1990s. The development of commerce and social, cultural and political links is contingent upon electronic modes of communication to a greater degree than ever before. The production, exchange and use of information increasingly depends upon access to the public network for voice, data or image communication. The widest participation of residential consumers and smaller and medium-sized businesses in this electronic environment can only be ensured if the appropriate technical, organizational and commercial characteristics of the public network are in place.

**Towards a Common Public Telecommunication Infrastructure**

The terms and conditions of access to the public network are only one, albeit an important, facet of the conditions needed for participation by the majority of the population in the electronic communication environment. If public policy regimes are unable to ensure that the necessary conditions are in place, the rate of exclusion from the electronic environment will grow, especially in regions or segments of the user population that are already disadvantaged in their ability to access the public network's traditional services. This trend will continue despite a profusion of innovative techniques of communication. It will result from the underlying political and economic forces of technical and institutional change that have been illustrated by the Strategic model of telecommunication development, and which characterize the political economy of industrial development in the 1990s.

The observations of another historian are salutary. Mumford argued that 'the problem of integrating the machine in society is not merely a matter of making social institutions keep step with the machine. The problem is also one of altering the nature and rhythm of machines to fit the actual needs of the community' (Mumford 1934: 367). The needs of the whole community of users of public telecommunication networks increasingly are being defined by the requirements of a relatively small number of large firms which seek advanced networks and services to augment their competitiveness in the international market. Yet, it must not be forgotten that the production of goods and services in global markets and the prospects of multinational firms depend upon the informed participation of workers and consumers and, therefore, upon their participation in the electronic communication environment.

If the regulatory process in Europe is to keep pace with technical innovations and the restructuring of telecommunication markets, globally, regionally and nationally, it will need actively to build a consensus on the conditions necessary for access to a common public telecommunication. Effective public intervention in the market will be as important in determining who is able to participate in the electronic communication environment of the twenty-first century as the competitiveness of the public network suppliers and the multinational network-using firms.

**Notes**

1 See Commission of the European Communities 1992a: 12, 1988a, 1990a; Council of the European Communities 1990a, 1990b, 1991, 1992.

2 Under the licensing procedure, providers of some competitive (non-reserved) services licensed in one of the European Community member states could request a Community-wide extension of that licence.

3 Subsidiarity refers to the fact that in all areas where it proposes action, the European Commission must consider whether the objectives of the proposed action cannot be sufficiently achieved by the member states or whether, by reason of the scale or efforts of the proposed

action, they can be better achieved by the Community (Commission of the European Communities 1992a). Currently, organizations include those such as the European Conference of Postal and Telecommunications Administrations (CEPT), which has a European Committee for Telecommunications Regulatory Affairs (ECTRA). The European Commission DG-XIII performs a range of functions which may reach the stage of agreement under the terms and conditions of Directives and other instruments available to the Commission under the Treaty of Rome. Two Committees are active, the ONP Committee and the Approvals Committee for Terminal Equipment (ACTE). There are also two advisory groups, the Senior Officials Group for Telecommunications (SOG-T) and the Joint Committee on Telecommunications Services. Relevant technical standards organizations include the European Telecommunications Standards Institute (ETSI), the European Committee for Standardization/European Committee for Electrotechnical Standardization (CEN–CENELEC) and the European Radiocommunications Committee – European Radiocommunications Office (ERC/ERO) which are under the auspices of CEPT. It is envisaged that all these organizations and groups would continue to operate if a CTC were created to play a co-ordinating role.

4 With respect to cost accounting systems for leased lines – (1) The costs of leased lines will include the direct costs incurred by the telecommunications organizations for setting up, operating and maintaining leased lines, and for marketing and billing of leased lines; (2) common costs, that is, costs which can neither be directly assigned to leased lines nor to other activities area allocated as follows: (a) whenever possible, common cost categories shall be allocated based upon direct analysis of the origin of the costs themselves; (b) when direct analysis is not possible, common cost categories shall be allocated based upon an indirect linkage to another cost category or group of cost categories for which a direct assignment or allocation is possible, and the indirect linkage shall be based on comparable cost structures; (c) when neither direct nor indirect measures of cost allocation can be found, the cost category shall be allocated based upon a general allocator computed by using the ratio of expenses directly assigned or allocated to, on the one hand, services which are provided under special or exclusive rights and, on the other, to other services (Commission of the European Communities 1992c: Article 10).

5 For example, in 1988 the Commission suggested that 'if reasonable overall cost-related tariffs are not achieved by 1st January 1992, the whole approach [to liberalization] with regard to the future evolution of the telecommunications sector will have to be re-evaluated' (Commission of the European Communities 1988b: 18).

6 Net Expenditure on Tangible Fixed Assets excluding customer premises expenditure in 1990 ECU.

7 These estimates cover an average penetration for telephone lines of forty-two per 100 population, customer lines connected to local digital exchanges equal to or greater than 75 percent, Integrated Services Digital Network access available to business subscribers in towns over 10,000 population, cellular mobile coverage for the Global System for Mobile Communication (GSM) reaching 90 percent of the population, and availability of Public Packet Switched Data services for all businesses. In 1989, the number of subscriber lines per 100 population varied in the European Community between forty-six per 100 population in Germany and France to eighteen per 100 population in Portugal. This amount is in addition to the Commission's expenditure of 770 million ECU between 1987 and 1991, and 200 million ECU from 1991 to 1993 (Commission of the European Communities 1992a).

8 Recourse cannot be made to these innovative technologies as a solution to the problems of network design. These new iterations are subject to similar political and economic contradictions as earlier generations of technology.

# Glossary

**accounting rate** A term referring to communications traffic between zones controlled by different PTOs. It is used for the establishment of international accounts and is expressed as a charge per traffic unit. The accounting rate differs from the collection rate, which is the charge to the telecommunication service user that is made by the PTO.

**AIN** Advanced Intelligent Network announced by Bellcore in 1989 to replace earlier versions of the Bellcore Intelligent Network (IN/1 and IN/2).

**analogue signal** Analogue electrical signal that directly represents another form of energy or activity such as the representation of sound waves in electrical form corresponding in frequency to the sound waves it represents. Contrasts with digital signals where information is expressed as a series of discrete numeric values.

**ATM** Asynchronous Transfer Mode is a technique for packet data transmission intended for fibre optic networks. In ATM each data unit (or cell) is of a fixed length.

**AT&T** American Telegraph and Telephone Company.

**automatic telephone exchange** An exchange or telephone switch that routes calls automatically to make the connections requested by calling parties without the intervention of an operator at an exchange.

**bandwidth** The bandwidth of a communication channel defines the range of frequencies that can be conveyed effectively in the channel.

**basic service** Definitions vary in different countries but the term generally refers to traditional telecommunication services associated with 'common carriage', 'conveyancing', and minimal information processing capabilities.

**B Channel** One of the communication channels carried on a digital transmission line between an exchange or central office and the ISDN subscriber. The 'B Channel' carries 64 kbit/s of user data as opposed to signalling information.

**Bellcore** Bell Communications Research Inc., the jointly-owned research arm of the RBOCs in the United States.

**bit-transporter** A bit-transporter is a PTO responsible for transmitting messages in the form of voice, data or image signals through a network without adding value by processing messages.

**BT** British Telecom.

**bundling** The components of intelligent networks that support network functionality can be bundled and unbundled in different configurations that affect the ways in which network functionality is made available for use by the network operator and/or users.

**bypass traffic** Bypass traffic may refer to traffic that originates or terminates without using the facilities of the PTO's exchanges, such as when a private network is used. It may refer to traffic that is routed using leased lines or microwave equipment that 'bypasses' or does not make use of the functionality in the public switched network.

**CCITT** International Telegraph and Telephone Consultative Committee of the ITU. Now the Telecommunication Standardization Bureau (TSB).

**CCSS7** Common Channel Signalling System No. 7 refers to an ITU CCITT recommendation for common channel signalling designed for use in networks conforming to Integrated Services Digital Network (ISDN), but is not limited to this configuration. This system provides internal control and network intelligence and defines the format and content of packets on the D Channel in the ISDN configuration.

**CELL** In a cellular radio network, a geographic area where subscribers are served by a radio base station.

**cellular network** A cellular radio network provides a communication service that uses mobile phones and radio waves as the transmission medium. A service provider's equipment switches the radio frequencies as a caller moves from one cell to another.

**CEPT** Conference of European Postal and Telecommunications Administrations.

**CLASS** Custom Local Area Signalling, a term used for advanced services for residential users in jurisdictions in the United States.

**CNET** Centre National d'Etudes des Télécommunications, France.

**common channel signalling** CCS, a method of providing control in a telecommunication network in which many traffic circuits may be controlled by a single pair of signalling channels. The route may contain several exchanges, each of which is equipped with centralized control to respond to the signalling functions.

**computerized switching technologies** Refers to data processing computer control techniques in the operation of a telecommunication switch or exchange. The technique is generally referred to as Stored Program Control (SPC).

**crossbar switch** Crossbar switches are electromechanical devices used in telephone exchanges to carry analogue signals. They consist of contacts arranged in a matrix, which are activated by a metal bar that is rotated by the action of a solenoid. These devices are constructed in large matrices to connect incoming and outgoing lines to establish calls in a circuit switched exchange.

**CTC** Community Telecommunications Committee, proposed by the Commission of the European Communities to achieve greater co-ordination of European telecommunication policies and directives.

**DACS** Digital Access Cross-connect Systems enable improved network management in a private network environment based upon leased lines.

**D Channel** The 'D Channel' carries control signals and low-speed data, and transmits at 16 or 64 kbit/s in the ISDN configuration.

**DEC** Digital Equipment Corporation.

**DGT** Direction Générale des Télécommunications, France.

**digital overlay networks** Networks that may or may not interwork with the public switched telephone network operated by the PTOs.

**digital signal** An electrical signal made up of discrete pulses coded to represent information. A digital switch is a device for making switched connections between circuits to establish transmission paths for digital data transmission and in which the connections are made by processing digital signals rather than analogue signals.

**DRG** Direction de la Réglementation Générale, France.

**duopoly** In the telecommunication field generally refers to the licensing of a second public switched telecommunication network operator in a domestic market, initially for a fixed period of time.

**ECU** European Currency Unit.

**ESS** Electronic Switching System referring to the AT&T family of switches.

**ETSI** European Telecommunications Standards Institute.

**FCC** Federal Communications Commission, United States.

**Feature Node** A name given to a node in a telecommunication network, which would incorporate enhanced intelligence and a greater degree of functionality offered as service features.

**front-end processor** A computer subsystem used mainly to interface a main computer or host processor to a communication network. It generally has responsibility for communication control rather than application programs.

**gateway** Equipment that connects a data communication network at one location to outside services and networks, and to networks at different locations. A gateway enables devices attached to a network to communicate across networks, such as public packet switched networks, and with other computers, such as within a corporate data centre that does not use the same communication protocols or procedures.

**gigabit-per-second switching** A switch capable of operating at speeds of thousands of millions ($\times 10^9$) of bits per second.

**hardwired** A component of a technology that is fixed at the time of manufacture or assembly such as soldered connections on printed circuit boards. Unlike software it cannot readily be changed.

**IBM** International Business Machines.

**IN/1, IN/2** The first and second versions of the Intelligent Network concept released by Bellcore prior to the announcement of AIN (Advanced Intelligent Network).

**insourced** Goods and services that are developed within an organization rather than purchased on the external market.

**institution** An institution can be defined as the embodiment of recognizable procedures that mediate the interactions among members of groups or collectivities within society, or in society as a whole.

**interface** A specification of the rules by which interaction between two separate functional units operate to conform with an overall system's requirements. An interface specification may include logical, electrical and mechanical specifications.

**INTUG** International Telecommunication User Group.

**ISDN** A set of standards defined by the ITU CCITT that defines a type of digital telecommunication service allowing the integrated transmission of voice, data and still pictures in digital form.

**ITU** International Telecommunication Union, Geneva.

**kbit/s, Mbit/s** Kilobits per second or megabits per second ($\times 10^6$) refer to units of data volume and are measures of the speed of transmission of digital data or bandwidth.

**LAN** Local Area Network.

**mainframe** A general term for a large centralized computer system.

**modem** Equipment used to link a digital device such as a computer or terminal to an analogue telephone line. The term is a contraction of modulator–demodulator, and its main function is to modulate an outgoing stream of digital data bits so that they are compatible with telephone networks designed to handle analogue traffic, and to reverse the process with an incoming bit stream.

**multiplexing system** A technique that permits the sharing of communication links among a number of user devices such as data terminals.

**NET** Norme Europeénne de Télécommunications. A mandatory standard established by the European Telecommunications Standards Institute.

**NETSTAR** Network Service Transaction and Recording System, a service management system introduced by AT&T.

**node** A general term meaning a point in a communication network at which several transmission lines meet. It is also used to refer to the equipment on which the lines terminate, and which controls and/or switches traffic. Telephone exchanges are nodes in a telephone network.

**OECD** Organisation for Economic Co-operation and Development, with headquarters in Paris.

**Oftel** Office of Telecommunications, the United Kingdom telecommunication regulatory body.

**ONA** Open Network Architecture, a US Federal Communications Commission mandate issued as part of the Computer III Inquiry.

**ONP** Open Network Provision, a Commission of the European Communities mandate issued as a Directive applicable to the member states.

**out-of-band signalling** A system of signalling in which signals are conveyed outside the band of frequencies normally used for message transmission but instead are carried on (or associated with) the transmission channel.

**outsourced** Goods or services purchased from an external or arms-length organization.

**PABX** Private Automated Branch Exchange.

**packet switching** In a circuit switched system a physical circuit is established between two terminals for the duration of a call. With high speed circuits carrying digital data, greater resource utilization can be achieved by sharing paths through the network. Packet switching allows data to be transferred through the network in packets, which include the data, addressing and sequencing information used to control the progress of the packet through the network.

**PBX** Private Branch Exchange.

**Primary Rate Interface** A configuration of channels used to transfer data and signalling information when accessing a network conforming to the Integrated Services Digital Network (ISDN) standard. In North America the interface operates at 1.544 Mbit/s, and uses twenty-three B Channels, at 64 kbit/s each, for information and one 'D Channel' at 64 kbit/s for signalling. In Europe the bit rate is 2.048 Mbit/s and uses thirty 'B Channels' plus one 'D Channel'.

**PSO** Private Service Operator.

**PTO** Public Telecommunication Operator.

**PTT** Post, Telegraph and Telephone Administration.

**Pulse Code Modulation** PCM is a method of converting an analogue signal into a digital signal which entails sampling the analogue signal and encoding it to represent the signal levels of the waveform of the analogue signal in digital form.

**RBOC** Regional Bell Operating Company.

**real-time** A system where information processing takes place immediately following the event that occasions it. Process control systems generally operate in real-time because they must process data arriving from the devices they are controlling fast enough to feed back information affecting their operation.

**SCP** Service Control Point in the intelligent network configuration.

**selector** An electromechanical switching device used in exchanges in which moving contacts are used to create connections for switching circuits.

**SMS** Service Management System in the intelligent network configuration.

**software-based functions** Functions that are embedded within, or peripheral to, the telecommunication network, which operate on the basis of software and computerized information processing.

**SPC** Stored Program Control enables the use of computer programs to achieve centralized control over the operational, administrative and maintenance functions of a telecommunication switch or exchange.

**SSCP** Service Switching and Control Point in the intelligent network configuration.

**SSP** Service Switching Point in the intelligent network configuration.

**STP** Signal Transfer Point in the intelligent network configuration.

**Strowger switch** A. B. Strowger developed a mechanical relay in 1891 and the principles are used in many countries as the basis of electromechanical switching operations in telephone exchanges.

**supplementary service** In the ISDN context, services offered in addition to the basic ISDN configuration such as sub-addressing.

**switch** Switching equipment directs communication traffic to alternative transmission lines such as telephone exchanges for voice traffic and packet switching exchanges for data traffic.

**TCAP** Transaction Capabilities Application Part provides the network transaction capability in CCSS7. The TCAP uses the system of Signal Transfer Points (STP) and is the basis for signalling for various intelligent network components.

**telematics services** Services that combine electronic technologies for collecting, storing, processing and communicating information. Such technologies may process information, such as computer systems, or they may disseminate information, such as a telecommunication network. Telematics is a term widely used in France. Terms such as informatics and information technology may also be used.

**terabit-per-second packet switch** A switch capable of operating at speeds of millions and millions of bits per second.

**transparent** Easily seen through, recognized or detected. A term adopted with increasing frequency by organizations such as the OECD and Commission of the European Communities to refer to the need for clear rules and procedures with respect to telecommunication standards, pricing, etc.

**T1** T1 refers to the transmission lines along which digital signals travel at a rate of 1.544 Mbit/s and to the common carriers' service that provides 1.544 Mbit/s transmission. In the United States, a T1 signal has the capacity of 24 voice channels. The equivalent data rate in Europe is 2.048 Mbit/s.

**value added service** Definitions vary in different countries. In the telecommunication field the term generally refers to more advanced telecommunication services associated with 'enhanced' or additional information processing capabilities not normally associated with 'basic' services such as voice telephony. Value added services are often candidates for supply in competitive, liberalized markets prior to consideration of liberalization of 'basic' service provision.

**videotex** Generic term for a two way information retrieval service using a specialized visual display terminal or a personal computer running special software. Public videotex services include the British Prestel and the French Minitel network. Various standards have been adopted in Europe, North America and Japan.

**virtual private network** VPN, a network using virtual circuits in a packet switching network where messages share links between various nodes using some form of multiplexing. There is no access path associated with each call to provide an end-to-end connection for the duration of the call. Messages are transmitted as a number of small packets of binary information.

**VSAT** Very Small Aperture Terminal used in connection with satellite services.

# Bibliography


Abel, G. (1988) 'Ericsson touts cutover of double-duty AXE switch'. *Communications Week*, 30 May.

Abernathy, H. L. and Utterback, J. M. (1978) 'Patterns of industrial innovation'. *Technology Review*, 80 (June/July): 2–29.

Abrahams, P. (1991) 'French group buys data network'. *Financial Times*, 8 June.

Ambrosch, W. D., Maher, A. and Sasscer, B. (eds) (1989) *The Intelligent Network: a Joint Study by Bell Atlantic, IBM and Siemens*. Berlin: Springer Verlag.

Analysys (1991) 'ONP: the Progress Report, European Telecommunications 2'. Cambridge: Analysys Publications, July.

Antonelli, C. (1985) 'The diffusion of an organisational innovation: international data, telecommunications and multinational industrial firms'. *International Journal of Industrial Organisation*, 3: 109–18.

Antonelli, C. (1988) *New Information Technology and Industrial Change: the Italian Case*. Dordrecht: Kluwer Academic Publishers.

Antonelli, C. (1991) *The International Diffusion of Advanced Telecommunications in Developing Countries*. Paris: OECD.

Armstrong, C. (1987) 'IBM press seminar', International Telecommunication Union, Telecom '87, Geneva, October.

Arnbak, J. C. (1987) 'Many voices, one structure: the challenge of telematics'. *The Information Society*, 5: 101–18.

Aronson, J. and Cowhey, P. (1988) *When Countries Talk: International Trade in Telecommunications Services*. Cambridge, MA: Ballinger.

Arrow, K. J. (1962) 'Economic welfare and the allocation of resources for inventions', in R. Nelson (ed.), *The Rate and Direction of Inventive Activity: Economic and Social Factors*. Princeton: Princeton University Press.

AT&T (1907) 'AT&T Annual Report'. American Telegraph & Telephone.

AT&T (1910) 'AT&T Annual Report'. American Telegraph & Telephone.

AT&T (1990) 'Intelligent networks – building network-based intelligent services'. *Trends in Telecommunications, AT&T Network Systems International*, October.

AT&T (1991) 'Towards the ultimate network'. AT&T Network Systems International, Hilversum, The Netherlands, June.

Babe, R. E. (1990) *Telecommunications in Canada*. Toronto: University of Toronto Press.

Bar, F. and Borrus, M. (1989) 'From public access to private connections: network strategies and competitive advantage in US telecommunications'. Berkeley Roundtable on the International Economy, OECD, Commission of the European Communities, Brussels, October.

Battarel, G., Kung, R., Martin, J. and Vilain, B. (1987) 'Introduction of new services in the French telephone network', International Switching Symposium, Phoenix.

Baumol, W. J., Panzar, J. C. and Willig, R. D. (1982) *Contestable Markets and the Theory of Industry Structure*. New York: Harcourt Brace Jovanovich.

Beau, O., Silva, J. and Verhille, H. (1990) 'Network aspects of broadband ISDN'. *Alcatel Electrical Communication*, 64 (2/3): 139–46.

Beesley, M., Laidlaw, B. and Gist, P. (1987) 'Prices and competition on voice telephony in the UK'. *Telecommunications Policy*, September, 230–36.

Begbie, R., Cunningham, I. and Williamson, H. (1982) 'INET: the Intelligent Network', International Conference on Computer Communications, London, 7–10 September.

Bell, D. (1973) *The Coming of Post-Industrial Society: a Venture in Social Forecasting*. New York: Basic Books.
Berben, C. (1988) 'Some remarks to intelligent networks', 10th International IDATE Conference, Montpellier, 16–18 November.
Berben, C. (1991) 'Regulatory Symposium, Forum '91', 6th World Telecommunication Forum, Geneva, 9–11 October.
Berger, P. and Luckman, T. (1966) *The Social Construction of Reality*. Chicago: Chicago University Press.
Bidal, G. and Mangin, P. (1991) 'Michel Feynerol, CNET – ATM: more than a trend'. *Télécom Réseaux International*, 49.
Blau, J. and Schenker, J. (1991) 'Europe assesses leased lines'. *Communications Week International*, 15 April.
Bohlin, E. and Granstrand, O. (1989) 'National monopolies in transition, strategic options for telecommunications administrations with illustrations from Sweden', International Telecommunications Society, European Regional Meeting, Budapest, 30 August.
Bonbright, J. C. (1962) *Principles of Public Utility Rates*. New York: Columbia University Press.
Bordeaux, D. and Maher, A. (1989) 'Intelligent networks – a European standard'. *Telecommunications International*, 67–9, February.
Braudel, F. (1984) *Civilization and Capitalism 15–18 Century*. Vol. III, *The Perspective of the World*. London: Collins.
Bregant, G. and Kung, R. (1990) 'Service creation for the intelligent network', International Switching Symposium, Stockholm.
Bressand, A. (1988) 'The age of synergy – telecommunications and trade'. *Intermedia*, 16 (1): 32–5.
Bressand, A. (1990) 'Perfection, and beyond: when markets become networks'. *Project Promethee Perspectives*, 13 (May): 7–15.
British Telecom (1989) 'Application programming interface: British Telecom/IBM study report'. British Telecom, IBM UK, London.
British Telecom (1990) 'News review'. British Telecom, London, August.
Bruce, R. R., Cunard, J. P. and Director, M. D. (1986) *From Telecommunications to Electronic Services: a Global Spectrum of Definitions, Boundary Lines, and Structures*. London: Butterworth.
BT (1991) 'BT launches Syncordia'. *BT Information Exchange*, October/November.
Carlsson, B. and Stankiewicz, R. (1991) 'On the nature, function and composition of technological systems'. *Journal of Evolutionary Economics*, 1: 93–118.
Carpentier, M., Farnoux-Toporkoff, S. and Garric, C. (1992) *Telecommunications in Transition*. Chichester: John Wiley & Sons.
Cawson, A., Morgan, K., Webber, D., Holmes, P. and Stevens, A. (1990) *Hostile Brothers: Competition and Closure in the European Electronics Industry*. Oxford: Clarendon Press.
Chandler, A. (1962) *Strategy and Structure: Chapters in the History of the Industrial Enterprise*. Cambridge, MA: MIT Press.
Chesnais, F. (1988) 'Technical co-operation agreements between firms'. *OECD STI Review*, 4 (December): 51–119.
Ciborra, C. U. (1992) 'Innovation, networks and organizational learning'. Pp. 91–102 in C. Antonelli (ed.), *The Economics of Information Networks*. Amsterdam: Elsevier Science Publishers.
Clark, J. M. (1961) *Competition as a Dynamic Process*. New York: The Brookings Institution.
Clark, K. (1987) 'Investment in new technology and competitive advantage'. Pp. 59–81 in D. Teece (ed.), *The Competitive Challenge: Strategies for Industrial Innovation and Renewal*. Cambridge, MA: Ballinger.
COGECOM (1991) *Rapport Annuel*, Groupe France Télécom, Paris.
Collet, P., Vilain, B., Fritz, P. and Cariou, J. (1984) 'On the short-term evolution of the French telephone network', International Switching Symposium Conference, Florence.
Commission of the European Communities (1983) 'Telecommunications (Communication

from the Commission to the Council)'. COM (83) 329 final, Commission of the European Communities, Brussels, 9 June.

Commission of the European Communities (1987) 'Towards a dynamic European economy: Green Paper on the development of a common market for telecommunication services and equipment'. COM (87) final, Commission of the European Communities, Brussels, 30 June.

Commission of the European Communities (1988a) 'Commission Directive of 16 May on competition in the markets in telecommunication terminal equipment'. 88/301/EEC, OJ L 131/73, Commission of the European Communities, Brussels, 16 May.

Commission of the European Communities (1988b) 'Towards a competitive community-wide telecommunications market in 1992: implementing the Green Paper on the development of a common market for telecommunications services and equipment'. COM (88) 48 final, Commission of the European Communities, Brussels, 9 February.

Commission of the European Communities (1990a) 'Commission Directive of 28 June 1990 on competition in the markets for telecommunications services'. 90/388/EEC, OJ L 192/10, 24.7.90, Commission of the European Communities, Brussels, 24 July.

Commission of the European Communities (1990b) 'Towards trans-European networks – for a Community action programme'. COM (90) 585 final, Commission of the European Communities, Brussels.

Commission of the European Communities (1991a) 'The European electronics and information technology industry: state of play, issues at stake and proposals for action'. Commission of the European Communities DG-XIII, Brussels.

Commission of the European Communities (1991b) 'Guidelines on the application of EEC competition rules in the telecommunications sector'. COM (91) C233/02 OJ C 233/2, Commission of the European Communities, Brussels, 6 September.

Commission of the European Communities (1992a) '1992 review of the situation in the telecommunications services sector'. Commission of the European Communities, Brussels, 21 October.

Commission of the European Communities (1992b) 'The European telecommunications equipment industry, the state of play, issues at stake and proposals for action'. Commission of the European Communities, Brussels, 25 June.

Commission of the European Communities (1992c) 'Proposal for a Council Directive on the mutual recognition of licences and other national authorizations for telecommunications services including the establishment of a single Community telecommunications licence and the setting up of a Community telecommunications committee'. ONPCOM91–79bis, COM(92) 254 final – SYN 438, Commission of the European Communities, Brussels, 3 March.

*Communications Daily* (1990) 'ONA seen as not working for answering service firms', n.d.

*Communications International* (1989) 'Intelligent network take off', August.

Communications Systems International (1988) 'European PTTs will find it difficult to operate one stop shopping'. *Communications Systems Worldwide*, May.

*Communications Week International* (1989) 'Untitled article', 13 November.

*Communications Week International* (1990) 'Partners selected by Bellcore, 1989–1990, cited in US Federal Register', 12 March.

*Communications Week International* (1991a) 'DBP Telekom plans IN upgrade', 4 April.

*Communications Week International* (1991b) 'Penetration of CCSS7', 4 November.

*Communications Week International* (1991c) 'Results of Yankee group survey', 4 February.

*Communications Week International* (1992) 'VPN Implementation in European Community Member States', based on Analysys data, 11 May.

Consultative Committee for International Telegraph and Telephone, International Telecommunication Union (1990) 'Intelligent network baseline document, liaison statement to working party XI/4 on intelligent network studies – the baseline document'. Document 404-E International Telecommunication Union, Geneva, 5–7, 14–23 March.

Coriat, B. (1989) 'The regulatory regime, market structure and business competitivity in France'. Berkeley Roundtable on the International Economy, OECD, Commission of the European Communities, Brussels, October.

Council of the European Communities (1986) 'Council Directive on the initial state of the mutual recognition of type approval for telecommunications terminal equipment'. 86/361/EEC OJ L 217/21, European Communities, Brussels, 24 July.
Council of the European Communities (1990a) 'Council Directive of 17 September 1990 on procurement procedures of entities operating in the water, energy, transport and telecommunications sectors'. 90/531/EEC, OJ L 297/1, 29.10.90, Council of the European Communities, 17 September.
Council of the European Communities (1990b) 'Council Directive of 28 June 1990 on the establishment of the internal market for telecommunication services through the implementation of Open Network Provision'. 90/387/EEC, OJ L 192/01, 24.07.90, Council of the European Communities, Brussels, 24 July.
Council of the European Communities (1991) 'Council Directive of 29 April 1991 on the approximation of the laws of the member states concerning telecommunications terminal equipment, including the mutual recognition of their conformity'. 91/263/EEC,OJ L 128/1, 23.05.91, Council of the European Communities, 29 April.
Council of the European Communities (1992) 'Council Directive of 5 June 1992 on the application of Open Network Provision to leased lines'. OJ L 165, 19.06.92, Council of the European Communities, 5 June.
Cronin, F. J., Parker, E. B., Colleran, E. K. and Gold, M. A. (1991) 'Telecommunications infrastructure and economic growth'. *Telecommunications Policy*, 15 (6): 529–35.
Cruise-O'Brien, R. (ed.) (1983) *Information, Economics and Power: the North–South Dimension*. London: Hodder & Stoughton.
Dang N'guyen, G. (1988) 'Telecommunications in France'. Pp. 131–54 in J. Foreman-Peck and J. Mueller (eds), *European Telecommunications Organisations*. Baden-Baden: Nomos.
Darmaros, T. (1992) 'Implementing the Integrated Services Digital Network (ISDN): prospects and problems in the realisation of a telecommunication concept'. DPhil dissertation, Science Policy Research Unit, University of Sussex.
*Data Communications* (1992) 'A VPN status report', June.
Dataquest Europe (1991) 'Market Statistics', London, November.
David, P. A. and Bunn, J. A. (1988) 'The economics of gateway technologies and network evolution: lessons from the electricity supply industry'. *Information Economics and Policy*, 3: 165–202.
David, P. A. and Greenstein, S. (1990) 'The economics of compatibility standards: an introduction to recent research'. *Economics of Innovation and New Technology*, 1 (1): 3–41.
Davies, A. (1991) 'The digital divide: a political economy of the restructuring of telecommunications'. DPhil dissertation, University of Sussex.
DBP Telekom (1991a) 'Communication networks made in Germany'. DBP Telekom, Bonn, October.
DBP Telekom (1991b) 'Report on the financial year, Telekom'. DBP Telekom, Bonn, August.
de Jonquieres, G. (1989) 'The deadly mirage of convergent technology'. *Financial Times*, 24 July.
de Sola Pool, I. (1990) *Technologies without Boundaries: On Telecommunications in a Global Age*, E. Noam (ed.). Cambridge, MA: Harvard University Press.
Debenham, R. (1989) 'Intelligent networks – a BT perspective'. BT Research Laboratories, Martlesham.
Demsetz, H. (1982) 'Barriers to entry'. *The American Economic Review*, 72: 47–57.
Department of Trade and Industry (1991) 'Competition and choice: telecommunications policy for the 1990s'. CM 1461, Department of Trade and Industry, London.
Department of Trade and Industry (1992) 'Intelligent networks: a consultative document from the Department of Trade and Industry'. Department of Trade and Industry, London, December.
Depouilly, B. (1990) 'Overview'. *Alcatel Electrical Communication* 64 (2/3): 114–15.
Dixon, H. (1990a) 'Hint of gloom in BT's price rise'. *Financial Times*, 19 July.
Dixon, H. (1990b) 'Oftel challenges phone rental charge rises'. *Financial Times*, 18 September.

Dorros, I. (1991) 'Two major forces driving the evolution of public switched networks', 6th World Telecommunication Forum, Technical Symposium, Geneva, 15 October.

Dosi, G. (1982) 'Technological paradigms and technological trajectories'. *Research Policy*, 11: 147–62.

Dosi, G. (1988) 'Sources, procedures and microeconomic effects of innovation'. *Journal of Economic Literature*, 26: 1120–71.

Dosi, G. and Orsenigo, L. (1988) 'Co-ordination and transformation: an overview of structures, behaviours and change in evolutionary environments'. Pp. 13–37 in G. Dosi, C. Freeman, R. Nelson, G. Silverberg and L. Soete (eds), *Technical Change and Economic Theory*. London: Pinter.

Dosi, G., Pavitt, K. and Soete, L. (1990) *The Economics of Technical Change and International Trade*. Hemel Hempstead: Wheatsheaf.

Dowling, M. J. and Ruefi, T. W. (1992) 'Technological innovation as a gateway to entry: the case of the telecommunication equipment industry'. *Research Policy*, 21: 63–77.

Duffy, A. (1992) 'GPT takes less than 50 per cent of BT digital switch orders'. *Telecom Markets*, 26 November.

Dupuis-Toubol, F. (1991) 'Deregulation or reregulation? France's Aggiornamento'. *Telecoms Réseaux International*.

Duysters, G. and Hagedoorn, J. (1992) 'Convergence and divergence in the international information technology industry', MERIT Conference, Convergence and Divergence in Economic Growth and Technical Change, Maastricht, 10–12 December.

Eburne, M. (1989) 'The British Telecom intelligent network programme'. British Telecom mimeo, London.

Eliot, T. S. (1969) 'The Hollow Men, 1925'. Pp. 83–6 in V. Eliot (ed.), *The Complete Poems and Plays of T. S. Eliot*. London: Guild Publishing.

Ellison, I. (1990) 'The telecommunications duopoly review: proposals for policy change'. London: Robert Fleming.

Ellison, I. (1991) 'The outstanding issues: a critical review', *Financial Times* Telecommunications Conference, London, 10 April.

Ergas, H. (1988) 'Learning from the trauma of telecoms today'. *Intermedia*, March: 26–9.

Ergas, H. (1992) 'France Télécom: has the model worked?' Seminar organized by the Royal Norwegian Council for Scientific and Industrial Research, 29 January.

Ericsson Ltd (1988) 'Network UK'. Ericsson Telecom, Brighton, Summer.

Ericsson Ltd (1990) 'The vital link'. Ericsson Telecom, Brighton, Summer.

Ericsson Telecom (1988) 'The intelligent network, 2, network architecture: open network strategies'. Ericsson Telecom.

Ericsson Telecom (1990) 'The intelligence behind network development'. Ericsson Telecom.

Eske-Crisstensen, B., Schreier, B. and Stroh, R. (1989) 'Intelligent network – a powerful basis for future services'. *Siemens Telecom Report*, Vol. 12 (5).

European Communities (1957) 'Treaty establishing the European Economic Community'. Signed at Rome, 25 March 1957.

European Communities (1987) 'Single European Act'. Official Journal of the European Communities, L169, 29 June.

European Court of Justice (1991) 'Judgment of the Court, Case C-202/88'. European Court of Justice, 19 March.

European Telecommunication Consultancy Organization (1990) 'Study on the interrelations between ONP concept and IN'. ETCO.60/3190.90, 2nd Interim Report, European Telecommunication Consultancy Organization, August.

European Telecommunications Standards Institute (1990) 'NA6 NA4'. Sophia Antipolis, February.

Evagora, A. (1991) 'ETSI scales back IN plans'. *Communications Week International*, 10 June.

Ewbank Preece Ltd (1993) 'Statistical research in telecommunications: disparities between the core and Less Favoured Regions of the EC Community, Final Report to the Commission of the European Communities'. Brighton, February.

Federal Communications Commission (1955) 'Hush-a-Phone Corp, order set aside, Hush-a-Phone Corp. v. United States, 238 F.2d 266 at 269 (DC Cir. 1956) on remand 22 FCC 112, 1957'. Washington, DC.
Federal Communications Commission (1959) 'Allocation of frequencies in the bands above 890 Mc'. Washington, DC.
Federal Communications Commission (1967) 'TELPAK tariff sharing provisions of American Tel. & Tel. Co. and Western Union Tel. Co., 8 FCC 2d 178'. Washington, DC.
Federal Communications Commission (1969a) 'Computer I, CC Docket 16979, First Report 17 FCC 2d 587'. Washington, DC.
Federal Communications Commission (1969b) 'Microwave Communications Inc., Nos. 16509–19'. Washington, DC, 13 August.
Federal Communications Commission (1971) 'Computer I, Final Decision 28 FCC 2d 267'. Washington, DC.
Federal Communications Commission (1979) 'Computer I, Tentative Decision 28 FCC 2d 291'. Washington, DC.
Federal Communications Commission (1981) 'Computer II, FCC Second Computer Inquiry, 77 FCC 2d 384 (1980), recon. 84 FCC 2d 50 (1980), further recon. 88 FCC 2d 512 (1981), aff'd sub. norm'. Washington, DC.
Federal Communications Commission (1983) 'BOC Separation Order, 95 FCC 2nd at 1130, FCC Report and Order, in re policy and rules concerning the furnishing of customer premises equipment, enhanced services and cellular communications services by the Bell Operating Companies, 95 FCC 2nd 1117 (1983) (BOC Separation Order), aff'd sub norm'. Washington, DC.
Federal Communications Commission (1984) 'In the matter of MTS and WATS market structure (Amendment of Part 67 of the Commission's rules and establishment of a Joint Board), *Decision and Order*, released 28 December'. Washington, DC, 28 December.
Federal Communications Commission (1987) 'Computer III, Report and Order, in amendment of sections 64.702 of the Commission's Rules and Regulations (Third Computer Inquiry) Docket No. 85–229, 104 FCC 2d 958 (1986) (Phase I Order), on reconsideration, 2 FCC RED 3035 (1987) (Phase I Reconsideration), 2 FCC RED 3072 (1987) (Phase II Order)'. Washington, DC.
Federal Communications Commission (1988) 'Phase I Order, Memorandum Opinion and Order on Further Reconsideration, in amendment of sections 64.702 of the Commission's Rules, 3 FCC RED 1135 (1988) (Phase I Further Reconsideration); and Order on reconsideration of Phase II Order, Memorandum Opinion and Order on Reconsideration, 3 FCC RED 1150 (1988) (Phase II Reconsideration)'. Washington, DC.
Federal Communications Commission (1990) 'Comments of the Washington Utilities and Transportation Commission, CC Docket No. 90–623, in the matter of Computer III Remand Proceedings, Bell Operating Company Safeguards and Tier 1 Local Exchange Company Safeguards'. Washington, DC, 7 March.
*Financial Times* (1989a) 'Brussels to rethink plans for telecoms deregulation', 28 April.
*Financial Times* (1989b) 'The EC muddle in telecoms', 22 April.
*Financial Times* (1989c) 'Into the era of the intelligent telephone', 19 April.
*Financial Times* (1989d) 'untitled article', 30 June.
Finnie, G. (1991) 'Network operators face their culture'. *Communications Week International*, 15 April.
Finnie, G. (1992) 'Software concerns dominate ISS'. *Communications Week International*, 9 November.
Fintech (1990) 'Mercury profits give rise to a ripple of investor confidence', based on Nomura Research and Henderson Crosthwaite data. *Telecom Markets*, 156, 28 June: 6.
France Télécom (1990) '1990 Annual Report'. France Télécom Group, Paris.
Frantzen, V., Maher, A. and Christensen, B. E. (1989) 'Towards the intgelligent ISDN concepts, applications, introductory steps', 1st International Conference on Intelligent Networks, Bordeaux, 14–17 March.
Freeman, C. (1982) *The Economics of Industrial Innovation*. London: Frances Pinter.

Freeman, C. (1988) 'Introduction'. Pp. 1–8 in G. Dosi, C. Freeman, R. Nelson, G. Silverberg and L. Soete (eds), *Technical Change and Economic Theory*. London: Pinter.
Freeman, C. and Perez, C. (1988) 'Structural crises of adjustment, business cycles and investment behaviour'. Pp. 38–66 in G. Dosi, C. Freeman, R. Nelson, G. Silverberg and L. Soete (eds), *Technical Change and Economic Theory*. London: Pinter.
Gabel, R. (1967) *The Development of Separations Principles in the Telephone Industry*. East Lansing: Michigan State University Press.
Gabel, R. (1969) 'The early competitive era in telephone communication, 1893–1920'. *Law and Contemporary Problems*, 34: 340–59.
Gabel, R. (1992) 'The impact of premium telephone services on the technical design, operation and cost of local exchange plant'. Working Paper C-30, Public Policy Institute, Washington, DC, January.
Gaffard, J-L. (1992) 'Untitled mimeo'. LATAPSES Sophia Antipolis, France, January.
Gannes, S. (1985) 'The judge who's reshaping the phone business'. *Fortune*, 1 April.
Garnett, G. W. (1985) *The Telephone Enterprise: the Evolution of the Bell System's Horizontal Structure, 1875–1900*. Baltimore: Johns Hopkins University Press.
Garnham, N. (1990) 'Telecommunications in the UK'. Discussion Paper No. 1, Fabian Society, London, October.
Garnham, N. and Mansell, R. (1991) *Universal Service and Rate Restructuring in Telecommunications*. OECD/ICCP Report No. 23, Paris: OECD.
GEC Plessey Telecommunications (1989) 'GPT Presentation', European Dataquest Conference, London.
Giacoletto, S. (1989) 'Planning and managing the enterprise communications infrastructure: the digital experience'. Digital Equipment Corporation International (Europe), Geneva, August.
Giddens, A. (1985) *The Nation-state and Violence: a Contemporary Critique of Historical Materialism*. Vol. 2. Cambridge: Polity Press.
Gilhooly, D. (1987) 'Towards the intelligent network'. *Telecommunications*, December.
Gilhooly, D. (1989a) 'The fight to level the open network playing field'. *Communications Week International*, 17 April.
Gilhooly, D. (1989b) 'France upgrades net'. *Communications Week International*, n.d.
Gilhooly, D. (1990) 'Nordic Services Group dissolved'. *Communications Week International*, 26 March.
Gilhooly, D. (1991a) 'Euro-watchdog'. *Communications Week International*, 4 November.
Gilhooly, D. (1991b) 'IN future at stake'. *Communications Week International*, 21 January.
Gilhooly, D. (1991c) 'Interconnect fees: the price of admission'. *Communications Week International*, 24 June.
Gillick, D. and Gilhooly, D. (1990) 'International users take command'. *Communications Week International*, 15 October.
Gist, P. (1990) 'The role of Oftel'. *Telecommunications Policy*, February: 26–52.
Goodhart, D. (1991a) 'Bonn eases telephone link rules'. *Financial Times*, 15 March.
Goodhart, D. (1991b) 'Private East German phone networks discovered'. *Financial Times*, 20 February.
Granstrand, O. and Sigurdson, J. (1985) 'The role of public procurement in technological innovation and industrial development in the telecommunication sector – the case of Sweden'. Pp. 147–85 in O. Granstrand and J. Sigurdson (eds), *Technological Innovation and Industrial Development in Telecommunications: the Role of Public Buying in the Telecommunications Sector in the Nordic Countries*. Gotenburg, Lund, Stockholm: Research Policy Institute, Lund University.
Granstrand, O. and Sjolander, S. (1990) 'Managing innovation in multi-technology corporations'. *Research Policy*, 19 (1): 35–60.
Gray, H. M. (1981) 'Reflections on Innis and institutional economics'. Pp. 99–110, in W. Melody, L. Salter and P. Heyer (eds), *Culture, Communication and Dependency: the Tradition of H. A. Innis*. Norwood, NJ: Ablex.

Haid, A. and Mueller, J. (1988) 'Telecommunications in the Federal Republic of Germany'. Pp. 155–80 in J. Foreman-Peck and J. Mueller (eds), *European Telecommunication Organisations*. Baden-Baden: Nomos.

Hamel, G., Doz, Y. L. and Prahalad, C. K. (1989) 'Collaborate with your competitors – and win'. *Harvard Business Review*, January – February: 133–39.

Handford, R. and Newman, M. (1992) 'DTI says BT can't keep its intelligence to itself'. *Fintech Telecom Markets*, 22 December.

Hawkins, R. (1992a) 'The doctrine of regionalism: a new dimension for international standardisation in telecommunications'. *Telecommunications Policy*. May/June: 339–53.

Hawkins, R. (1992b) 'Standards for technologies of communication: policy implications of the dialogue between technical and non-technical factors'. Science Policy Research Unit, University of Sussex, DPhil dissertation.

Hayes, D. (1990) 'VPNs in Europe'. *Communications Week International*, 10 December.

Henderson, R. M. and Clark, K. B. (1990) 'Architectural innovation: the reconfiguration of existing product technologies and the future of established firms'. *Administrative Science Quarterly*, 35: 9–30.

Heuermann, A. (1987) 'Der Market fuer Mahrwetdienste in der Bundesrepublik Deutschland'. Working Paper Report No. 25, Wissenschaftliches Institut für Kommunikationsdienste, Bad Honnef, February.

Hiergeist, F. (1991) 'Telekom 2000: the construction programme for the telecommunications network in eastern Germany', European Telecommunications Industry Conference, London, 7–8 November.

Hills, J. (1986) *Deregulating Telecoms: Competition and Control in the United States, Japan and Britain*. Westport, CT: Quorum Books.

Hodgson, G. (1988) *Economics and Institutions: a Manifesto for a Modern Institutional Economics*. Cambridge: Polity Press.

Horowitz, R. (1989) *The Irony of Regulatory Reform: the Deregulation of American Telecommunications*. Oxford: Oxford University Press.

Hu, Y. S. (1992) 'Global or transnational corporations and national firms with international operations'. *California Management Review*, 34 (2): 107–27.

Huber, P. (1987) 'The geodesic network: 1987 report on competition in the telephone industry for the Department of Justice in accordance with the Court's decision in the Matter of US v. Western Electric Company, 552 Supp. 131, 194–5, DDC 1982, January'. Department of Justice, Washington, DC, January.

Independent Commission for World-wide Telecommunication Development (1984) 'The missing link: report of the Independent Commission for World-wide Telecommunication Development'. International Telecommunication Union, Geneva.

Innis, H. A. (1951) *The Bias of Communication*. Toronto: University of Toronto Press.

International Resource Development (1990) 'Network management hardware, software and services markets'. *Communication International.*

International Switching Symposium (1990) 'France Télécom's Intelligent Network', Mimeo, Stockholm.

International Telecommunication Union Advisory Group on Telecommunication Policy (1989) 'The changing telecommunication environment: policy considerations for members of the ITU'. International Telecommunication Union, Geneva, February.

Irwin, M. R. (1984) *Telecommunications America: Markets Without Boundaries*. Westport, CT: Quorum Books.

Irwin, M. R. and Niman, N. B. (1988) 'The corporate telecommunications network: market transparency and state accountability', Communications Policy Research Conference, Windsor Great Park, 22 June.

Jacobson, R. (1989) *An 'Open' Approach to Information Policy Making: a Case Study of the Moore Universal Telephone Service Act*. Norwood, NJ: Ablex.

Jhally, S. (1990) *The Codes of Advertising: Fetishism and the Political Economy of Meaning in the Consumer Society*. New York: Routledge.

Jouet, J., Flichy, P. and Beaud, P. (eds) (1991) *European Telematics: the Emerging Economy of Words*. Amsterdam: North Holland.

Kay, M., Whalley, S. and Smith, N. (1989) 'ROSA-RACE Open Systems Architecture'. BT Research Laboratories mimeo, Martlesham.

Killette, K. (1991) 'US losing network lead'. *Communications Week International*, 4 November.

Konig, K. (1989) 'Intelligent networks in the context of the Community's telecommunications policy', 1st International Conference on Intelligent Networks, Bordeaux, 14–17 March.

Korzeniowski, P. (1988a) 'Partners to be part of IBM's future'. *Communications Week*, 28 March.

Korzeniowski, P. (1988b) 'Users to see SS7 benefits'. *Communications Week*, 30 May.

Kung, R. (1989) 'Key design issues in the intelligent network', 1st International Conference on Intelligent Networks, Bordeaux, 14–17 March.

Kung, R., Martin, J., Collet, P. and Lapierre, M. (1990) 'Introduction of France Télécom's intelligent network', International Switching Symposium, Stockholm.

Lanvin, B. (1991) 'The dialogues that could have been'. *Project Promethee Perspectives*, 15 (January): 9–18.

Leadbeater, C. (1990) 'Marriages of convenience'. *Financial Times*, 29 May.

Leiss, W. (1976) *The Limits to Satisfaction*. London: Marion Boyars.

Lundvall, B-A. (1989) 'Innovation, the organised market and the productivity slow-down', OECD Seminar on Science, Technology and Economic Growth, Paris, 5–8 June.

Lynch, K. (1991a) 'International resale ready to run'. *Communications Week International*, 18 February.

Lynch, K. (1991b) 'Operators band against breakdowns'. *Communications Week International*, 4 November.

Lynch, K. (1991c) 'SS7 timebomb ticks'. *Communications Week International*, 2 September.

Lynch, K. and Herman, E. (1991) 'Outages continue to plague US'. *Communications Week International*, 18 November.

Magedanz, T. and Popescu-Zeletin, R. (1991) 'Modelling open network provision and intelligent networks'. Pp. 29–42, in W. Effelsberg, H. W. Meuer and G. Muller (eds), *Kommunikation in Verteilten Systemen*. Berlin: Springer Verlag.

Maher, A. T. (1989) 'The intelligent network – a European standard?' *Telecommunications*, February.

Maher, A. T., Christensen, B. E. and Tebes, J. (1988) 'The intelligent network learning curve', IEEE Global Telecommunications Conference, Florida, 28 November.

Mallinson, K. (1992) 'Europe's "CLASSless" Society'. *Communications Week International*, 9 November.

Mansell, R. (1982) 'The "new dominant paradigm" in communication: transformation versus adaptation'. *Canadian Journal of Communication*. 8 (3): 42–60.

Mansell, R. (1986) 'The telecommunications bypass threat: real or imagined?' *Journal of Economic Issues*, March: 145–64.

Mansell, R. (1989) *Telecommunication Network-based Services: Policy Implications*, ICCP OECD Report No. 18. Paris: OECD.

Mansell, R. (1990) 'Rethinking the telecommunications infrastructure: the new "black box"'. *Research Policy*. 19: 501–15.

Mansell, R. and Hawkins, R. (1992) 'Old roads and new signposts: trade policy objectives in telecommunication standards'. Pp. 45–54, in P. Slaa and F. Klaver (eds), *Telecommunication: New Signposts to Old Roads*. Amsterdam: IOS Press.

Mansell, R. and Jenkins, M. (1992) 'Networks and policy: interfaces, theories and research'. *Communications & Strategies*. 1[er] trimestre (5): 31–50.

Mansell, R. and Sayers, D. (1992) 'European cross-border telecommunication: the large business user's view'. Science Policy Research Unit, University of Sussex, Brighton, October.

Mansell, R., Holmes, P. and Morgan, K. (1990) 'European integration and telecommunications: restructuring markets and institutions'. *Prometheus*. 8 (1): 50–66.

Martin-Lof, J. (1989) 'Sweden policy and organisation within the telecommunication sector'. Corporate Planning, International Affairs, Televerket, Stockholm, 21 February.

Melody, W. H. (1977) 'Telecommunications jurisdictional cost separations: the US experience as a guide to Canadian policy'. Philadelphia: University of Pennsylvania, 23 February.

Melody, W. H. (1985) 'The information society: implications for economic institutions and market theory'. *Journal of Economic Issues* XIX (2): 523–39.

Melody, W. H. (1986) 'Telecommunication: policy directions for the technology and information services'. *Oxford Surveys in Information Technology*, 3: 77–106.

Melody, W. H. (1987a) 'Information: an emerging dimension of institutional analysis'. *Journal of Economic Issues*, XXI (3): 1313–39.

Melody, W. H. (1987b) 'New policy options: price caps, social contracts and flexible pricing – do they make sense and can they substitute for rate of return?' in W. Bolter (ed), *Policy Symposium on Federal/State Price-of-Service Regulation: Why, What and How?* George Washington University, Washington, DC.

Melody, W. H. (1989) 'Efficiency and social policy in telecommunication: lessons from the US experience'. *Journal of Economic Issues*, XXIII (3): 657–88.

Melody, W. H., Salter, L. and Heyer, P. (eds) (1981) *Culture, Communication and Dependency: the Tradition of H. A. Innis*, Norwood, NJ: Ablex.

Metcalfe, J. S. and Boden, M. (1992) 'Evolutionary epistemology and the nature of technology strategy'. Pp. 49–71 in R. Coombs, P. Saviotti and V. Walsh (eds), *Technology Change and Company Strategies: Economic and Sociological Perspectives*. London: Harcourt Brace Jovanovich.

Metcalfe, J. S. and Reeve, N. (1990) 'On technological taxonomy'. Mimeo, Manchester: PREST University of Manchester, March.

Meurling, J. and Jeans, R. (1985) *A Switch in Time: an Engineer's Tale*. Chicago: Telephony Publishing Corporation.

Miles, R. E. and Snow, C. C. (1986) 'Organisations: new concepts for new forms'. *California Management Review*, XXVIII (3): 62–73.

Ministère des Postes et Télécommunications (1992) 'Annual report'. Direction de la Réglementation Générale, Paris, March 91–March 92.

Molina, A. H. (1989) *The Transputer Constituency: Building up UK/European Capabilities in Information Technology*. Edinburgh: Research Centre for Social Sciences, University of Edinburgh.

Morgan, K. (1989) 'Telecom strategies in Britain and France: the scope and limits of neo-liberalism and dirigisme'. Pp. 19–55, in M. Sharp and P. Holmes (eds), *Strategies for New Technology*. London: Philip Allan.

Mudd, D. R. and Starkey, M. Z. (1992) 'Economic implications of collocation for the exchange access market'. *Telecommunications Policy*, 16 (6): 511–26.

Mueller, J. and Foreman-Peck, J. (1988) *Liberalising European Telecommunications*. Cambridge: Blackwell.

Mulgan, G. J. (1991) *Communication and Control: Network and the New Economies of Communication*. London: Guildford Press.

Mumford, L. (1934) *Technics and Civilization*. London: Routledge & Kegan Paul.

National Telecommunications and Information Administration (1991) 'The NTIA Infrastructure Report: Telecommunications in an Age of Information'. United States, Department of Commerce, Washington, DC, October.

Nelson, R. and Winter, S. (1982) *An Evolutionary Theory of Economic Change*. Cambridge, MA: Belknap Press.

Neumann, K-H. (1986) 'Economic policy toward telecommunications, information and the media in West Germany'. Pp. 131–52, in M. Snow (ed.), *Marketplace for Telecommunications*. New York: Longman.

Neumann, K-H. (1990) 'The unification of telecommunications in Germany', Brussels Meeting of the International Institute of Communications Telecommunications Forum, Brussels, 16–17 July.

Neumann, K-H. and von Weizsaecker, C. (1982) 'Tariff policy for leased circuits and private branch exchanges from the economic point of view'. Pp. 145–89, in Deutsche Bundespost (ed.), *1982 Yearbook of the DBP, Bonn*. Bonn: Deutsche Bundespost.

Noam, E. (1987) 'International telecommunications in transition', Conference on Communications and Technology, Brookings Institution, New York, 9 May.

Noam, E. M. (1992) 'Beyond the golden age of the public network'. Pp. 6–10, in H. Sapolsky, R. Crane, W. Neuman and E. Noam (eds), *The Telecommunications Revolution*. London: Routledge.

Noble, D. F. (1977) *America by Design: Science, Technology and the Rise of Corporate Capitalism*. New York: Alfred A. Knopf.

Noble, D. F. (1986) *Forces of Production: a Social History of Industrial Automation*. New York: Oxford University Press.

Noll, R. and Owen, B. (1987) 'United States v. AT&T: an interim assessment'. Studies in Industry Economics, Discussion Paper No. 139. Stanford University, California, August.

Nordling, O. (1990) 'Perspective on political processes evolving from the new telepolicy in Europe – the Swedish approach', INDC 90 Conference, Farsta, Sweden.

Northern Telecom (1987) 'Annual Report'.

Northern Telecom (1988) 'Annual Report'.

Northern Telecom (1989) 'Annual Report'.

Norton, J. (1990) 'Threats and opportunities in pan-European communications'. *Telecommunications*, March.

OECD (1988) *The Telecommunications Industry: the Challenges of Structural Change*, OECD ICCP Report No. 14. Paris: OECD.

OECD (1992) 'The 1992/3 communications outlook'. OECD, Paris, 14 August.

Ohmae, K. (1990) *The Borderless World*. London: Collins.

Onians, F. A. (1989) 'A view of the intelligent network'. *International Congress on Business, Public and Home Communications, EuroComm '88*. Amsterdam: North Holland, pp. 125–45.

Oram, R. (1989) 'US giant plugs in its global ambitions'. *Financial Times*, 30 August.

Owen, D. (1989) 'No room for wrong numbers'. *Financial Times*, 20 March.

Pacific Bell (1987) 'The intelligent network task force report'. Pacific Bell Task Force, Pacific Telesis, San Francisco, CA, October.

Pacific Bell (1988) 'Pacific Bell's response to the intelligent network task force report'. Pacific Telesis, San Francisco, CA.

Perez, C. (1983) 'Structural change and the assimilation of new technologies in the economic and social system'. *Futures*, 15 (4): 357–75.

Pierce, M., Fromm, F. and Fink, F. (1988) 'Impact of the intelligent network on the capacity of network elements'. *IEEE Communications Magazine*, December.

Pisano, G. P., Russo, M. V. and Teece, D. J. (1988) 'Joint ventures and collaborative arrangements in the telecommunications equipment industry'. Pp. 23–70, in D. C. Mowery (ed.), *International Collaborative Ventures in US Manufacturing*. Cambridge, MA: Ballinger.

Porat, M. U. and Rubin, M. R. (1977) *The Information Economy*, 9 vols. Washington, DC: Department of Commerce, Government Printing Office.

Porter, M. E. (1990) *The Competitive Advantage of Nations*. New York: Free Press.

Porter, M. E. (1992) 'On thinking about deregulation and competition'. Pp. 39–44, in H. Sapolsky, R. Crane, W. Neuman and E. Noam (eds), *The Telecommunications Revolution*. London: Routledge.

Porter, M. E. and Fuller, M. B. (1986) 'Coalitions and global strategy'. Pp. 315–43, in M. E. Porter (ed.), *Competition in Global Industries*. Boston: Harvard Business School Press.

Posner, R. A. (1969) 'Natural monopoly and its regulation'. *Stanford Law Review*. 21 (February): 548–643.

Pospischil, R. (1988) 'Ansatze zur Neuroganisation des Franzosischen Fernmeldewesens'. Discussion Paper No. 39, Wissenschaftliches Institut für Kommunikationsdienste: Bad Honnef.

Price Waterhouse (1991) 'Information technology 1991/92'. London.

PTT Telecom (1991) 'Annual Review'. PTT Telecom (the Netherlands).

Ralph, B. (1988) 'Virtual private systems from a public network', Which Way Forward for Corporate Telecommunication Networks, IBC Conference, London, 6 December.

Ravaioli, P. (1991) 'Necessity of both competition and cooperation', Regulatory Symposium, Forum '91, International Telecommunication Union, Geneva, 9–11 October.

Reddy, N. M. (1990) 'Product self-regulation: a paradox of technology policy'. *Technological Forecasting and Social Change*, 38: 49–63.

Regan, E. (1988) 'Telecom liberalisation and electronic banking'. *Transnational Data and Communications Report*, August/September.

Reich, R. B. (1991) *The Work of Nations*. New York: Vintage Books.

Rhodes, T. (1991) 'Deregulation: a Mercury perspective', 13th International IDATE Conference, Montpellier, 22 November.

Richardson, G. B. (1972) 'The organisation of industry'. *Economic Journal*, 82 (837–1532): 883–96.

Ricke, H. (1991) 'Telekom on the global market'. *Funkschau, Special*, 4 October.

Robinson, J. and Eatwell, J. (1973) *An Introduction to Modern Economics*. London: McGraw-Hill.

Robinson, P. (1991) 'Globalization, telecommunications and trade'. *Futures*, October: 801–14.

Robotham, J. and Walko, J. (1989) 'The Nordic connection'. *Communications Systems World-wide*, December/January.

Robrock, R. B. (1991) 'The intelligent network – changing the face of telecommunications'. *Proceedings of the IEEE*, 79: 7–20.

Rogerson, D. (1989) 'Cost-based tariffs for data services: position and trends in Europe'. OVUM Consultants, London.

Rosario, M. and Schmidt, S. K. (1991) 'Standardisation in the European Community: an example of ICT'. Pp. 183–96, in C. Freeman, M. Sharp and W. Walker (eds), *Technology and the Future of Europe: Global Competition and the Environment in the 1990s*. London: Pinter Publishers.

Rosenberg, G. B. (1982) *Inside the Black Box: Technology and Economics*. Cambridge: Cambridge University Press.

Rutkowski, A. M. (1988) 'The global information fabric'. *Project Promethee Perspectives*, 6 (June).

Samarajiva, R. and Shields, P. (1992) 'Emergent institutions of the "intelligent network": toward a theoretical understanding'. *Media, Culture and Society*, 14 (3): 397–419.

Samuel, J. (1989) 'AT&T woos potential tariff 12 users'. *Communications Week*, 1 May.

Saunders, R. J., Warford, J. G. and Wellenius, B. (1983) *Telecommunications and Economic Development*. Baltimore: World Bank and Johns Hopkins University Press.

Schenker, J. (1989) 'DEC study criticizes nets'. *Communications Week*, 26 September.

Schenker, J. (1990) 'German angst: user protests surcharges'. *Communications Week International*, 29 January.

Schenker, J. (1991) 'Searching for a single solution'. *Communications Week International*, 13 May.

Schenker, J. (1993) 'VANS market (OVUM London)'. *Communications Week International*, 5 April.

Scherer, F. M. (1980) *Industrial Market Structure and Economic Performance*. Chicago: Rand McNally.

Scherer, F. M. and Perlman, M. (1992) 'Introduction'. Pp. 1–10, in F. M. Scherer and M. Perlman (eds), *Entrepreneurship, Technological Innovation, and Economic Growth: Studies in the Schumpeterian Tradition*. Ann Arbor: University of Michigan Press.

Scherer, J. (1991) 'European telecommunications law'. Pp. 225–42, in A. Meijboom and C. Prins (eds), *The Law of Information Technology in Europe 1992*. Deventer-Boston: Kluwer Law and Taxation Publishers.

Schmidt, S. (1991) 'Taking the long road to liberalization: telecommunications reform in the Federal Republic of Germany'. *Telecommunications Policy*, 15 (3): 209–22.

Schoenbauer, H. (1993) 'Experience of the Deutsche Bundespost IN trial', IEE 'The intelligent network – fact or fantasy?' Tutorial seminar, Radcliffe House Conference Centre, near Coventry, 16 February.

Schumpeter, J. A. (1943) *Capitalism, Socialism and Democracy*, London: Allen & Unwin.

Schumpeter, J. A. (1954) *A History of Economic Analysis*. E. B. Schumpeter (ed.). New York: Oxford University Press.

Selwyn, L. L. and Montgomery, W. P. (1987) 'A report to the Ad Hoc Telecommunications Users Committee and the International Communications Association'. Economics and Technology Inc., Boston, 13 March.

Setchell, A. (1989) 'Computing aspects of intelligent networks'. Digital Equipment Corporation, Geneva, May, [41–22] 709.4789.

Shepherd, W. (1969) 'Communications: regulation, innovation and the changing margin of competition'. *Technological Change in the Regulated Industries*, Brookings Institution, New York, 6–7 February.

Shepherd, W. G. (1984) '"Contestability" vs. competition'. *American Economic Review*, 74 (September): 572–85.

Siemens (1990) 'The intelligent integrated broadband network – telecommunications in the 1990s'. Siemens mimeo.

Slaa, P. (1988) 'ISDN as a design problem: the case of the Netherlands'. Netherlands Office of Technology Assessment, Amsterdam, March.

Sleath, C. (1993) 'Intelligent network – fact or fantasy?' IEE Tutorial Seminar with the technical assistance of GPT, Radcliffe House Conference Centre near Coventry, 16 February.

Smythe, D. W. (1957) *The Structure and Policy of Electronic Communications*. Chicago: University of Illinois Bulletin.

Smythe, D. W. (1972) 'Reflections on proposals for an international programme of communications research', International Association for Mass Communication Research, General Assembly on Communication and Development, Buenos Aires, September.

Smythe, D. W. (1978) 'The political character of science (including communication science) or science is not ecumenical'. Pp. 171–76, in A. Mattelart and S. Siegelaub (eds), *Communication and Class Struggle: 1 Capitalism, Imperialism*. New York: International General Publishers.

Smythe, D. W. (1979) 'Realism in the arts and sciences'. Pp. 98–111, in K. Nordenstreng and H. Schiller (eds), *National Sovereignty and International Communication*. Norwood, NJ: Ablex.

Smythe, D. W. (1984) 'New directions for critical communications research'. *Media, Culture and Society*, 6: 205–17.

Smythe, D. W. and Dinh, T. V. (1983) 'On critical and administrative research: a new critical analysis'. *Journal of Communication*, 33 (3): 117–27.

Soderberg, L. (1991) 'The intelligent network develops'. *Telecommunications*, April.

Soete, L. (1991a) 'Technology and economy in a changing world', OECD Conference on Technology and the Global Economy, Montreal, 3–6 February.

Soete, L. (1991b) 'Technology in a changing world, OECD technology economy programme – policy synthesis'. MERIT, University of Limburg, the Netherlands, January.

Solomon, R. J. (1991) 'New paradigms for future standards'. *Communications & Strategies*, 2: 51–90.

Svedberg, B. (1989) 'Trends within public telecommunications digital networks'. *Telecommunications Journal*, 56 (IX): 588–92.

Sweeney, T. (1989) 'Northern, IBM team up'. *Communications Week*, 6 November.

Sweeney, T. (1991) 'Software problems snag public nets'. *Communications Week International*, 15 July.

Tahim, K. (1990) 'Regulation of telecommunications tariffs in the UK, British Telecom', Fundesco Conference, Madrid, January.

Taylor, R. (1990) 'Swedish Telecoms to be privatised'. *Financial Times*, 23 September.

Technology Investment Partners and Palo Alto Management Group (1990) 'Intelligent networking in the US: status and trends', Report for Commission of the European Communities and European Telecommunications Consultancy Organisation, 23 May.

Teece, D. J. (1989) 'Innovation, cooperation and antitrust: balancing competition and cooperation'. *High Technology Law Journal*, 4 (1): 1–131.

Telecommunication Managers Association (1993) 'The intelligent network: the TMA response to the Department of Trade and Industry Consultative Document'. Telecommunications Managers Association, Orpington, Kent, April.

Televerket (1990) 'Recommendations for a change in the operating conditions of Swedish Telecom (Televerket), Swedish Telecom proposal to the Minister of Transport and Communications'. Televerket, Farsta, Sweden, 14 December.

Temin, P. and Galambos, L. (1987) *The Fall of the Bell System: a Study in Prices and Politics*. Cambridge: Cambridge University Press.

Thomas, G. (1992) 'Cross border telecommunications: the prospects for pan-European carriers'. Science Policy Research Unit, University of Sussex, Brighton, April.

Thomas, G. and Miles, I. (1990) *Telematics in Transition: the Development of New Interactive Services in the United Kingdom*. Harlow: Longman.

Thorelli, H. B. (1986) 'Networks: between markets and hierarchies'. *Strategic Management Journal*, 7: 37–51.

Thorn Ericsson (1987) 'AXE in the UK'. Autumn.

Thorngren, B. (1990) 'The Swedish road to liberalisation'. *Telecommunications Policy*, April: 94–7.

Touche Ross Management Consultants (1991) 'Telecommunication tariffs trends in the European Community 1980–1990, Executive Report, Final Report to the Commission of the European Communities'. London, December.

Trebing, H. M. (1969a) 'Common carrier regulation – the silent crisis'. *Law and Contemporary Problems*, 34: 299–329.

Trebing, H. M. (1969b) 'Government regulation and modern capitalism'. *Journal of Economic Issues*, 3 (1): 87–109.

Trebing, H. M. and Melody, W. (1969) 'Staff Papers Report No. 1, an evaluation of domestic communications pricing practices and policies'. President's Task Force on Communications Policy – the Domestic Telecommunications Carrier Industry, Washington, DC.

Ungerer, H. and Costello, N. (1988) *Telecommunications in Europe*. Brussels: Commission of the European Communities.

United States (1976) 'Citation reference to Mumm v. Illinois, 1876'. Supreme Court, 94 US 113, 1976.

United States (1980) 'MCI v. AT&T, 74 C 633'. United States District Court of the Northern District of Illinois, Eastern Division, 13 June.

United States (1982a) 'United States v. AT&T, CA No. 74-1698, 82-0192 (DDC), Judge Greene, Order'. United States District Court of the District of Columbia, 24 August.

United States (1982b) 'United States v. AT&T, CA No. 82-0192 (DDC) Modification of Final Judgment'. United States District Court of the District of Columbia, 24 August.

United States (1983a) 'Computer and Communications Industry Assoc. v. FCC, 693 F 2d 198 (DC–Dir. 1982), Cert. denied, 1035 Ct. 2109 (1983)'. United States District Court.

United States (1983b) 'United States v. AT&T. 552 F. Supp. 131 (DDC 1982), aff'd., 460 US 1001 (1983)'. United States District Court of the District of Columbia.

United States (1985) 'Illinois Bell Tel. Co. v. FCC, 740 F.2d 465 (7th Cir. 1984), recon. denied, 49 Fed. Reg. 26.056 (June 26, 1984), aff'd. sub norm, North American Telecommunications Ass'n v. FCC, 772 F.2d 1282 (7th Cir. 1985)'. United States 7th Circuit Court.

United States (1987) 'United States v. AT&T, CA No. 92-0192 (DDC) Report and Recommendations of the United States Concerning the Line of Business Restrictions Imposed on the Bell Operating Companies by the Modification of Final Judgment'. United States District Court of the District of Columbia, 2 February.

United States (1990a) 'California v. FCC 905 F2d. 1217'. United States Court of Appeals, 9th Circuit.

United States (1990b) 'People of the State of California v. FCC, 905 F.2d 1217 (9th Cir. 1990)'. United States, 9th Circuit Court.

Vallance, I. (1990a) 'British Telecom's policy on cross subsidies'. *Financial Times*, 26 July.

Vallance, I. (1990b) 'Telephone rates and the UK review'. *Financial Times*, 5 April.

Vallance, I. (1991) 'BT'. *Financial Times*, 12 April.

Vallance, I. (1993) 'Competition and regulation in the global telecommunications industry', Speech to the Cranfield School of Management, Cranfield, 13 January.

Viesti, G. (1988) 'International cooperative agreements: new strategies for international growth and technological learning', 14th Annual Meeting of the European International Business Association, Berlin, 11–13 December.

von Weizsaecker, C. (1987) 'The economics of value added network services'. Report for IBM, Cologne.

Waverman, L. (1990) 'R&D and preferred supplier relationships: the growth of Northern Telecom', International Telecommunications Society Conference, Venice, March.

Webb, S. (1987) 'Sweden to open telecom's market'. *Financial Times*, 11 June.

Wieland, B. (1988) 'Current trends in telecommunications policy'. *Intermedia*, 14 (6): 13–8.

Williams, R. (1977) *Marxism and Literature*. Oxford: Oxford University Press.

Williamson, O. E. (1975) *Markets and Hierarchies: Analysis and Antitrust Implications*. New York: Free Press.

Williamson, O. E. (1985) *The Economic Institutions of Capitalism*. New York: Free Press.

Witte, E. (1987) *Neuordnung der Telekommunikation, Bericht der Regierungskommission Fernmeld*. Heidelberg: R v Decker's Verlag.

# Index